---

systems engineering series

Linear Systems Theory
Ferenc Szidarovszky, University of Arizona
A. Terry Bahill, University of Arizona

Engineering Modeling and Design
William L. Chapman, Hughes Aircraft Company
A. Terry Bahill, University of Arizona
A. Wayne Wymore, Systems Analysis and Design Systems

Model-Based Systems Engineering
A. Wayne Wymore, Systems Analysis and Design Systems

The Theory and Applications of Iteration Methods
Ioannis K. Argyros, Cameron University
Ferenc Szidarovszky, University of Arizona

System Integration
Jeffrey O. Grady, Jeffrey O. Grady System Engineering

system integration

Jeffrey O. Grady

CRC Press
Boca Raton London New York Washington, D.C.

Library of Congress Cataloging-in-Publication Data

Grady, Jeffrey O.
System integration / Jeffrey O. Grady.
p. cm. — (Sytems engineering series)
Includes index.
ISBN 0-8493-7831-1
1. Systems engineering. 2. System analysis. 3. Concurrent engineering.
I. Title. II. Series.
TA168.G65 1994
620'.001'1—dc20 94-5639
CIP

International Standard Book Number 0-8493-7831-1
Library of Congress Card Number 94-5639
Printed in the United States of America 4 5 6 7 8 9 0
Printed on acid-free paper

Contents

Preface

This book was conceived as a companion to an earlier work titled *System Requirements Analysis* (SRA). These two activities, SRA and integration, form twin buttresses for the system engineering process connected through the creative design process for the many items that comprise a system. Yet there are few books available on these subjects particularly as they are being influenced by the concurrent development notion. I was stimulated to start work on this book after a conversation with a colleague, Mr. G.D.H. (Dan) Cunha, who had volunteered to teach a course on this subject at the University of California San Diego Extension systems engineering certificate program. Dan could find no adequate textbook and we had some difficulty determining the boundaries and focus of the subject.

Later, work on a generic system development process diagram at General Dynamics Space Systems Division yielded the insight that several very different kinds of integration work were needed in a concurrent engineering environment. Some integration work had to be done in a cross-functional and co-product fashion while other work aligned with a co-functional and cross-product orientation. By applying a common system engineering thought process (a search for completeness) to these insights, it evolved that one could use a simple three-space coordinate system to express all integration situations in terms of points, lines, planes, and spaces.

With this aid, one could be sure that all of the possibilities would be covered. It also appeared that it would be possible to apply this theoretical construct in a practical way to help engineers understand the full sweep of integration situations. That is, it would be possible to teach people how to do system integration work in a classroom. Thus was born a book to fill the void that Dan and I had earlier uncovered.

Throughout the writing of this book I tried very hard to detune the content to recognize the widest possible range of customers from Department of Defense (DoD) to commercial in all product fields. I confess this was difficult because my principal experience had been in DoD programs and systems. There is an increasing interest, I know, in applying the systems approach in the non-DoD government and commercial sectors, and several colleagues in the National Council on Systems Engineering (NCOSE) were helpful in sensitizing me to these efforts. Doubtless, I am not completely cured and the reader interested in commercial applications may have to tolerate an appeal to what some would consider unnecessary rigor in some parts of the book. There is, however, another side to this question and that is that the DoD-derived process has a great deal to offer to everyone.

Author

Jeffrey O. Grady is the Principal of Jeffrey O. Grady System Engineering, a consulting firm focusing on helping firms achieve excellence in the systematic development of complex products through methodological improvements and education. He previously worked in industry for over 25 years in the capacities of engineering manager, system engineer, project engineer, field engineer, and customer training instructor. He has worked in several aerospace companies on a wide range of systems including space transport, cruise missiles, unmanned aircraft, underwater fire control, and superconducting magnets with a customer base that included the DoD (USAF and USN), NASA, DoE, and private corporations. Mr. Grady also served in the U.S. Marines in the communications field. He received a M.S. degree in systems management from the University of Southern California and a B.A. in mathematics from San Diego State University. In addition to this book, Mr. Grady has authored *System Requirements Analysis,* McGraw-Hill, 1993, and *System Engineering Planning and Enterprise Identity,* soon to be published by CRC Press, Inc. He was the first elected Secretary of the National Council on Systems Engineering (NCOSE) and the founding Editor-in-Chief of *Systems Engineering, The Journal of the NCOSE.* Mr. Grady teaches system engineering courses at the University of California, San Diego, and the University of California, Irvine, in the system engineering certificate programs he helped to found.

Acknowledgments

This book would not have been written without the encouragement of my wife, Jane. Shortly after finishing a previous book, I expressed to her that it would take too long to write a book on integration, the other part of the system engineering story. She reminded me how fast things seem to come together once you start. And so it is. Thank you, Jane.

Many thanks to Mr. Bernard Morais, President of Synergistic Applications Inc., and Mr. Jim Lacy, President of Jim Lacy Consulting for their helpful suggestions as reviewers of an earlier version of this manuscript. I also benefitted from conversations with Mr. Gene Northup, Executive VP of Leading Edge Engineering, Inc., who shined a light in the direction of the future of the system engineering profession unconstrained by the boundaries of our facilities through advances in computer technology.

I most appreciate the discussion in 1991 with Mr. Dan Cunha, then of General Dynamics Convair Division, about the difficulty in developing a strategy for teaching this subject for it was this stimulus that started me thinking thoughts that now appear here.

McGraw-Hill, the publisher of *System Requirements Analysis* (SRA), has given its approval for my use of sections from that book in this book and that is very much appreciated. The sections of the SRA book used in this book are: 7.4.2 (Chapter 12 of this book), Interface Analysis; 9.4.1 (7.4), Generic Specialty Engineering Process; 10.3.1 (13.4.2), Requirements Set Attributes; and 10.3.2 (13.4.2), Individual Requirements Attributes. Many thanks to McGraw-Hill for this courtesy.

We are the sum of every story we have ever heard or read. I have heard many, many system engineering stories, very few of them of a humorous nature, from my colleagues in the field of system engineering over more than a few years, especially in the past several since the founding of the National Council on Systems Engineering (NCOSE). After a while one cannot associate the sources with the specific stories nor even discriminate a story from it's aggregate effect on our current knowledge base. Paraphrasing Willy Nelson, let me say, "To all the system engineers I have ever known, thanks for the effect you have had on me, no matter your intent at the time." I hope the content of this book, some of which came from you at some time in the past, will have a positive effect on you and your future.

JOG December 1993
San Diego, CA

System Integration

Part I

The fundamentals

chapter one

Introduction

1.1 Integration. What is it?

One of the most often used words, yet most neglected notions, in the application of the system engineering process in industry is the word *integration*. It is a word like *system*. It has so many meanings and shades of gray that the listener or reader is never quite sure exactly how another person is using the word. Whatever integration is, however, it is a universally accepted necessity in the development of complex systems. It is at the downstream end of an organized process of decomposing complex problems into many smaller related problems, solution of the smaller problems by teams of specialists, and integration of the results into a solution to the original larger problem.

The system integration process is the art and science of facilitating the marketplace of ideas that connects the many separate solutions into a system solution. System integration is a component of the system engineering process that unifies the product components and the process components into a whole. It ensures that the hardware, software, and human system components will interact to achieve the system purpose or satisfy the customer's need. It is the machinery for what some call concurrent development.

The general approach in system engineering involves decomposing a customer need into actions or functions that the system must satisfy in order to satisfy that need. These lower tier functional requirements are then allocated or assigned to specific things or components in the system. These allocated functions are translated into performance requirements and combined with design constraints to form a set of minimum attributes (requirements) that a design team must satisfy in order for the component to fit into the overall system solution.

These requirements are synthesized by a team of engineers into one or more design concepts and, if more than one, the alternatives are traded one against the other to select a preferred solution. The design concept is expanded into a preliminary and detailed design interspersed with reviews to assure the team is on a sound path to success and in agreement with the work of other teams. Specialists from many disciplines work with the principal designer or design team to assure that the solution accounts for specialty requirements with which the designer may not be fully familiar.

Designs are translated into manufacturing planning and procurement decisions and production and procurement work begun. Test article versions of the product elements are subjected to special testing to qualify them for use in solution of the customer's need.

1.2 Toward a more effective process

But in this process, is there not some way to characterize the activity called integration so that we may all agree on its meaning and upon which we can base a more effective and foolproof deployment of its power into our own system development programs? In this book we establish a premise that integration has a finite number of component parts, each uniquely definable, and follow the consequences of this premise to their ultimate conclusions. We will decompose integration work into its fundamental parts and integrate the results. Yes, we are going to apply the system engineering process to improving our understanding of this word and concept.

This book is a companion work to *System Requirements Analysis*, written by the same author, which covers the decomposition and requirements identification process occurring on the front end or down stroke of development programs. Integration work principally occurs on the tail end or up stroke of development programs. These two activities are separated by the creative design process that synthesizes requirements into design solutions which must be integrated into a system solution. These two system engineering methodologies, together with verification by testing and analysis, comprise the heart and soul of the systems approach to development of complex systems.

Some observers of the process expressed in these two books have concluded that it is only appropriate to apply it to grand systems with a Department of Defense (DoD) customer known for its deep pockets. The author believes the system engineering process is perfectly applicable to smaller scale developments as well as large ones and those oriented toward commercial customers as well as DoD and NASA customers. The Mars mission space transport system will include within it an on-board computer (many actually, but certainly one) that can be treated as a system by its supplier development team. That computer will include a power supply that could be treated as a system by its vendor design team. A company that develops and manufactures video recorders for the commercial market could apply these same principles as could a company whose product line is bean bags.

Some of the component parts of the process may be overkill for some combinations of product lines (like bean bags), customer bases, and degrees of difficulty in the development process or maturity of the corresponding technology needed. The skillful development team will select, or the experienced customer will insist on, the components that make sense for the particular program.

It is true that the process, in the past, has been primarily applied by the aerospace industry for large government customers interested in very complex systems. Much of the language used to express system engineering ideas and references for parts of the process are derived from this historical reality. The author has made a conscious effort to present the system integration process from a generic perspective but is himself a captive of the same history and may not have succeeded completely.

There is another side to this story, however. The author believes that people in firms in the commercial marketplace are little different from those in the DoD industry in their resistance to integrated and structured development. The systems approach to problem solving has been available for many years yet there are many managers, directors, and vice presidents in engineering organizations that support DoD and NASA customers who distrust it and avoid its successful application.

It is true that there are aerospace firms with very fine system engineeirng capabilities that they effectively apply with good results. The stories are many, however, and in agreement with some of the author's own experiences in industry, of frustrated attempts to implement an effective system engineering process. The story goes something like this:

a. The customer will require in a request for proposal some evidence that a company has a system engineering capability. Typically this may take the form of a requirement that a system engineering management plan (SEMP) be submitted with the proposal. The company will write a credible SEMP with no intent to organize or work as described therein once the contract is awarded. Note that under MIL-STD-499A, which was the source of the SEMP requirement, the SEMP was not contractually binding.
b. When the contract is awarded to this firm, work begins using autonomous design groups interacting on an ad hoc basis. The customer becomes concerned that the company has neither a system engineering organization nor process and encourages the company to comply with their process requirements stated in the statement of work and the SEMP prepared by the contractor.
c. Finally, to get the customer off their backs after the design solution has been essentially determined, the company creates a system engineering function and assigns personnel to the program to accomplish a parallel system engineering process along the lines described in the SEMP. This team may be isolated from the design people and even told not to interact with them. They produce the standard set of system engineering products that may entail differences from the predetermined design solution and these products are essentially ignored.
d. At the earliest possible moment when the customer has been minimally satisfied that a systems approach has been applied, this group is disposed of.

Despite this history, even in some of our finest companies, some tremendous products have been developed that have served their customers (DoD and the general public) well. In some cases, not all, we will never know how much less these products would have cost or what other features or capabilities they may have had if a sound systems approach had been applied in their development. There appears to be a natural resistance to the systems approach driven by a human urge for power and control by those in functional management and by a human urge to belong to a group by the members of these organizations. The resulting autonomy of the functional organizations is the enemy of the systems approach on programs and it is alive and well in more places doing DoD work than one would think possible after decades of DoD insistence on an effective systems capability.

The point is that this same resistance, inspired by the same human motives, exists in commercial firms as well and prevents them from crossing the bridge sincerely looking for an effective organized product development approach which DoD has inspired but not been fully successful in obtaining from its contractors. Implementing an effective systems capability requires leadership from the top to impress everyone with the understanding that product success and a happy customer take priority over employee or management attitudes and company organizational dynamics. Product success results from customer value and that can be encouraged through an organized approach to problem solving called system engineering.

The author believes that the systems approach and happy employees are not mutually exclusive. The organizational structure and techniques encouraged in this book will provide an environment within which people, design engineers as well as system engineers, can excel in their profession. It will result in the imposition of the minimum set of constraints on the solution space within which the design community must seek a solution. It will result in the lowest cost and greatest value to the customer in the product they receive. Commercial companies that introduce appropriate elements of this process and educate their employees to perform them well will have an advantage over their competitors in terms of product cost, time to market, and product performance.

As we depart from a phase of history characterized by intense global military competition requiring increasingly sophisticated military systems, we will find that complex systems will be associated with a broader range of needs than in the past and the pace of development will be retarded from that of the past driven by a survival instinct. There will be a greater need to re-engineer old systems to upgrade their capabilities than to develop totally new systems pushing the state of the art in several directions at once. There will be a greater appeal to precedented systems than unprecedented ones involving a completely clean sheet of paper. Regardless of these changes, however, we are unlikely to discover a method for developing these new and modified systems that is more effective that the methods discussed in this book for they are based on the way we humans function.

1.3 Book organization

The book is composed of three parts. Part I, Chapters 1 through 5, explores the fundamentals of integration and offers a suggested organizational structure used throughout the book in other discussions. Part II, Chapters 6 through 10, focuses on process integration for the organization accomplishing a development program. This includes program planning, program tracking, discontinuity management, and organizing for concurrent development. Part III, Chapters 11 through 16, deals with product integration.

Part II is the province of system engineers who look at their profession as a management science the application of which allows them to get complex things done. People who do this work are commonly called project engineers and they are responsible for planning programs, setting up budgets, defining planned work in statements of work, and tracking programs to the planned budget and schedule. Part III is tuned to the system engineers who view their profession as an engineer solving complex product-oriented problems. These engineers work at the product interfaces and seek out inconsistencies between the system requirements and the current concept or design and within the overall work product of several different development teams.

System integration embraces both of these perspectives, process and product integration. While it is not uncommon for people with very different experiences to perform these two different functions on actual programs, the essence of the work they perform has a great deal in common. They work in the same patterns with different subject matter. Also, recent U.S. Air Force work on a concept called integrated management system has helped to more intimately connect these two component parts of the integration picture than heretofore and we will take advantage of that work in this book. In addition to isolated product and process integration, we must accomplish integration between these two entities as well in order to achieve the best possible product and customer value. Our reputation and reward will flow from how well we produce sound value for our customers. Excellence in system integration is a necessary prerequisite to the development of effective systems that customers will find useful, effective, and affordable.

We begin in this part with an initial definition of integration followed by four chapters that build a sound foundation in the fundamentals. We must answer the question, "What is integration and why is integration necessary?" As we will find in Chapter 2, the need for integration is driven by human limitations and competition because these same forces encourage specialization and decomposition of complex problems. Chapter 3 offers what the author claims is an optimum organizational structure for performing system integration work. This is, at least, the one that will be assumed in the book. Chapter 4 breaks the overall integration problem up into 27 simple integration components, each of which can be easily understood and applied. Why are there 27, and not 25 or 13, and how can we be sure there are no more? We can be sure because we will apply a sound systems approach that ensures completeness in our search for integration components. Since integration is nine-tenths human communications, understanding the relationship of integration work to information systems is essential. We will find in Chapter 5 that a big part of the communications answer is available in the computer systems that many companies have already paid for. But, how can we use these tools more effectively?

In Part II (Chapters 6 through 10), the book applies a subset of the 27 integration components to the development process, to the system that gives birth to the product system, to the system of which we humans are a part. Chapter 6 explains how a recent U.S. Air Force initiative called the integrated management system can be applied by any company in any product field to satisfy customer needs and improve contractor efficiency through development of sound planning that is clearly related to customer needs. In Chapter 7, we discuss the principal integration techniques for program execution and the greatest challenge in industry today, how do we organize the work of many specialists to be brought to bear on common problems without destroying their creativity? When a program is executed in accordance with the plan and everything goes according to that plan (Chapter 8), there is very little new to discuss not covered elsewhere in the book, so this is a short chapter. This is the proper realm of management. Chapter 9 deals with the more common situation of program execution under faulted conditions leading to discontinuities that must be healed. We must understand the problem, re-plan the remaining program, and get back on the plan. Finally, in Chapter 10, we will explore ways to channel our best efforts toward making us better hitters every time we step up to the plate.

In order to perform well in product integration, the topic of Part III, we must have satisfied all of the prerequisites covered in Parts I and II. We must have a sound integration capability at the program level embedded within our way of organizing the humans before we can hope to approach excellence in product integration. Chapter 11 begins the product integration discussion with a comprehensive methodology for ensuring we fully understand what is in the system and who or what organization is responsible for its development. This knowledge is linked to topics covered in Part II with a map between our product organization and the human organization.

The study of product interface possibilities in Chapter 12, largely derived from the book *System Requirements Analysis*, provides insights into how we should organize ourselves to minimize the problems at these interfaces. Most system engineers would tell you that if you are going to have problems in the development of a system it is going to occur at certain interfaces. At which interfaces will the problems occur and is this predictable? And, how can we minimize these problems? Chapter 13 covers integration during the requirements definition period.

Chapters 14, 15, and 16 bring all of the previous material together in the principal product system development arenas of design, test, analysis, production, and operation.

1.4 System development process overview

1.4.1 Process models

Several models of a structured, top-down system development process have been offered and three are summarized here. The author used the first one discussed below as his model but content of this book is applicable to any of these three models for they are but different views of the same process.

1.4.1.1 Traditional model

The first model is illustrated by a simple generic process diagram from which tasks are extracted and applied to the work called for on a given program phase. DoD and NASA programs are accomplished in program phases to control the cost risk between major review and decision events. Figure 1-1 illustrates this general process. Goals are established for each phase and work to date is reviewed at the end of each phase against those goals. A subset of the complete generic development process tasks are accomplished in one phase to achieve these goals. On the next phase, which may be preceded by a competitive proposal cycle, some of these tasks are repeated and some new tasks are accomplished while moving closer to program development completion.

Iteration occurs in this process for two reasons. First, while working from the top down, we become much better educated about the higher system levels as we work our way through lower tier solutions. This requires some re-work of prior conclusions at higher level. Also, we iterate our work in order to move deeper into the system hierarchy. Many of the same tasks required to characterize a subsystem, for example, are then required to characterize the major elements of that subsystem and are repeated for each of these elements. And so this process goes down to the components. The generic process diagram portrays a comprehensive set of sequential tasks needed to fulfill system engineering goals while the phasing diagram gives the programmatic and contractual view of the world. This is the model used by the author in this book but the others can be applied.

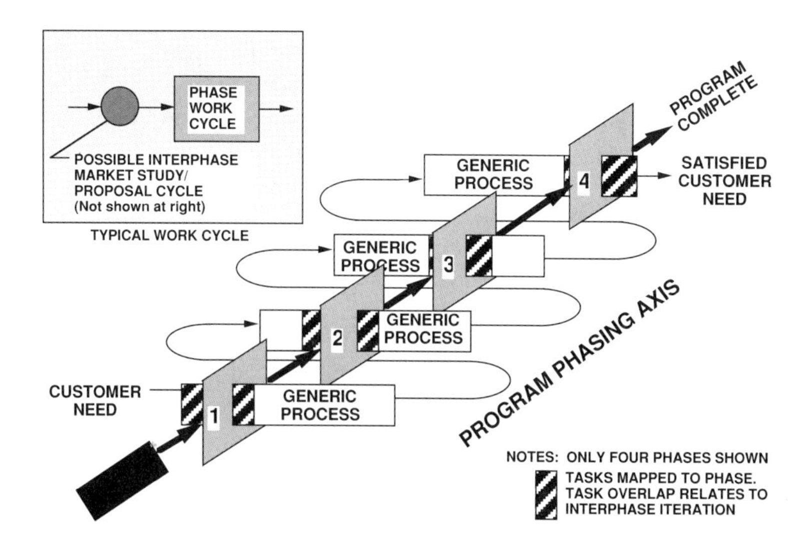

Figure 1-1 Traditional model of development.

1.4.1.2 The "V" model

The Center For Systems Management in Santa Clara, California, in the persons of Kevin Forsberg and Harold Mooz, has popularized a process diagram formed into a "V" shape with a decomposition and definition down stroke on the left and an integration and verification up stroke on the right. Figure 1-2 shows the general nature of this concept but it does not illustrate the detailed system architecture levels uncovered on the down stroke and the correspondingly layered integration and verification process on the upstroke characteristic of that model.

1.4.1.3 The spiral model

The spiral model, first adapted to software development by Barry W. Boehm;, presents an image of iterative cycling between several fundamental development activities while working toward more detailed knowledge of the preferred solution to the customer's need. Figure 1-3 offers a simplified variation on this model. While the original intent of this model followed a predictable pattern, some persons with a commercial interest see this as a useful model of reality where they get off the spiral when the product is good enough to first market within the context of a competitive situation. Others take this one step further and think in terms of a continuous spiral over time releasing market-ready products, with increasingly noteworthy features, flowing to market.

1.4.1.4 A common desire

In all of these views of the system development process, we seek to illustrate a multi-dimensional phenomena in two dimensions on paper and it is very difficult. We work to develop systems from the top down for reasons explained in Chapter 2. We begin with relatively little knowledge of the details and work toward complete knowledge of the solution to the problem expressed by the customer need. There is an experimental aspect to this process in that you must make decisions without sure knowledge as you work your way into a condition of more knowledge and some of these decisions will later be shown to have been less than the best of choices.

There is a need, therefore, for iteration in the system development process. At the same time, it is economical to apply essentially the same development process for all elements of the system. Since the system is composed of many things in a hierarchy and we work from the top down, we will be repeating one process over and over again in time. So, the same process occurs in time repeatedly in an expansive pattern during decomposition and definition and in a retracting pattern during integration and verification. This combination of cyclical action and the unidirectional nature of time make it difficult to portray the process clearly. The accompanying illustrations are three imperfect examples of attempts to do so.

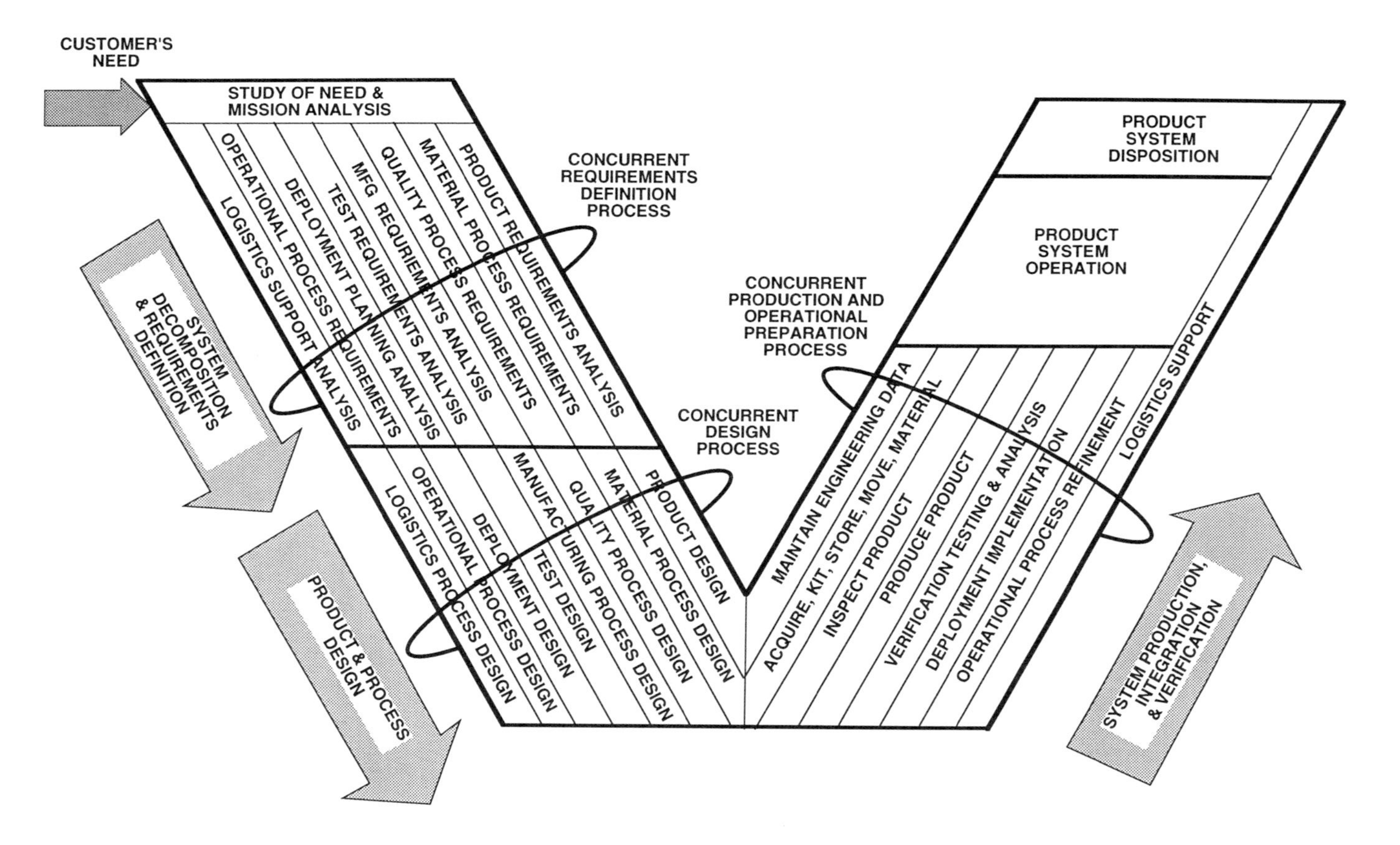

Figure 1-2 The "V" model of system development.

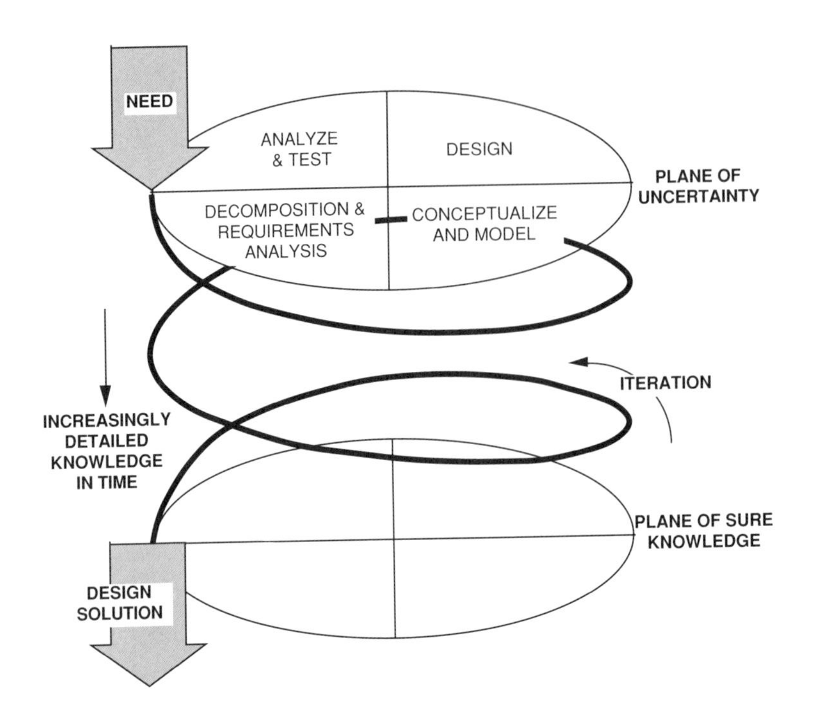

Figure 1-3 The spiral model of system development.

1.4.2 Process definition

Process diagrams are used throughout to provide a specific, documented process description commonly called for in the ISO 9000 series and MIL-STD-499B. We make the point repeatedly throughout the book that a company should have a standard, written process description telling how it accomplishes the system engineering process. This document might be called a system engineering manual (SEM) or system engineering management plan (SEMP). If your company has an effective, documented system engineering process derived over the years from trial and error and conscious throught to improve it based on program experiences, you should consider yourselves fortunate. The author salutes you.

If you do not have such a history and documentation, you should start now on that road while there is still time. The process diagrams included in this book may be useful as a basis for a similar set created by you for your company. The author has also written a book titled *System Engineering Planning and Enterprise Identity* that includes a generic SEM/SEMP embodying the principles discussed in this book. It includes a complete set of these process diagrams and accompanying database defining each task depicted by the blocks of the process diagrams. If you do not have such a written manual or plan, you may wish to use the one included in that book as a basis for preparation of one. The text may be acquired in computer word processor and graphics form yielding to editing at your facility for the special features of your situation.

Much of the content of this book has been consciously coordinated with MIL-STD-499B because that standard will influence how we think about this process for a long time to come. This will be true whether we are defense practitioners or commercial ones. In the former case, we will be directly influenced because we will have to comply with it in our process definition. In the latter case, we will be indirectly influenced by it because no one has done a better job of defining an effective system engineering process than DoD. DoD has, after all, had an opportunity over a very long time to observe the ultimate effects of good and bad processes on complex product qualities.

A lot of good thinking has gone into MIL-STD-499B by many very talented system engineers. The author admittedly is not in complete agreement with all of its content and such is the case with any grand design. So, the company plan should be coordinated with MIL-STD-499B tailoring needed to make it compatible with the company's process. Many fear the introduction of MIL-STD-499B as the end of their business. They feel this way not because of the content of the standard but because its implementation will require them to have an effective system engineering process and they do not have one. The new standard is an example of a description of a sound process. Your process may differ but you should be able to describe it in writing. If you have not yet done so, the content of MIL-STD-499B is not a bad place to begin.

Commercial practitioners may first view the contents of MIL-STD-499B as overkill, excessively costly, damaging to time-to-market needs, and stifling of creativity. Members of the design community in some defense firms where the systems process has only been given lip service will have similar feelings. It is very hard to accept that a sound system engineering job on a program will reduce program cost, improve program schedule, and result in greater customer value in the product. The fundamental problem in reaching this conclusion is that panic too often sets in at the beginning when time must be allotted to truly understanding the problem before leaping to design solutions. When we later have to re-design the product a thousand times (several thousand engineering changes after the design is complete is not all that uncommon on large defense programs) based on information derived out of problems exposed in testing and manufacturing, we strangely and foolishly accept that as part of our normal way of doing business.

Commercial practitioners may wish to steer toward a condition of fewer formal phases than commonly applied on DoD programs and more intense iteration volatility than a DoD practitioner would be comfortable with. But all will realize benefits from an organized process for the coordinated development of product and process described here.

A common concern expressed by design engineers in companies dealing in a DoD as well as a commercial product line is that "imposition" of a systems approach will stamp out creativity. While this has happened before, when the process has been implemented by zealots with no appreciation for the need for balance between order and creativity, it represents a poorly implemented systems approach. Perhaps the most extreme example of this exists in the imposition of the Communist state on the peasants of Russia for

70 years. State planning was accomplished at the highest level with insufficient information about lower tier need. No closed loop system existed like our market system to encourage adjustments leading to improved performance. Rigid prescriptions and a lack of any reward for new ideas destroyed creativity. In this book we seek to strike the optimum balance between order and creativity, far removed from this situation, that protects the development team from bad choices while not screening them from great opportunities.

Some system engineering fanatics will be disappointed that the author has not named this book systems integration. Some system engineers seem to feel that the word system must be used in its plural form when referring to system engineering and all of its components. Heated arguments derive from this small matter. It may make sense to use the plural form, as in systems engineering or systems integration, where one functional organization must deal with the development of systems on two or more programs. But, some other fanatics, including the author, prefer to cling to the singular form for no particular reason other than it just seems to sound right. If you prefer, you may read the plural form throughout the book without loss of meaning.

chapter two

The human basis for integration

2.1 Human limitations drive integration

The need to integrate the work of two or more people stems from some very fundamental principles of system engineering, human psychology, and human existence. Large and complex systems are composed of components that appeal to many different technologies because they must satisfy very complex needs that will not yield to simple solutions. We humans are limited in the amount of information, knowledge, and technology we can master and the amount of knowledge that we have created and provided access to far exceeds our individual limitations. There is an economic benefit realized by those who can take full advantage of available knowledge since they will be able to solve more complex problems than those who have mastered a smaller knowledge base. As the world turns, mankind finds new and more complex challenges to conquer encouraging more complex combinations of available technology.

The universal solution to the mismatch between man's capacity and available knowledge is for individual engineers to specialize in narrow fields and find ways to pool the talents of a team of specialists from engineering, quality, manufacturing, procurement, and other fields to form one equivalent super engineer. This solution is the basis for the birth of system engineering as a process, career, and, in some companies, an organization. It is also the basis for the need for requirements analysis and integration, the twin foundation stones of the system engineering process.

The specialization solution to this problem is thousands of years old. The time is long past, by several thousand years, when one human could master available knowledge. In very early societies, men specialized in hunting or gathering probably based on their individual physical abilities. Men with excellent eyesight, instincts, and physical ability were much better at hunting game then and have excelled in hunting their fellow man in jet fighters more recently. In medieval Europe, blacksmiths and tinsmiths had very different techniques. The industrial revolution accelerated this trend rapidly. But, the information explosion made possible by the computer and military competitions of the 20th century have dramatically encouraged the evolution of more finely specialized engineers and ushered in the need for effective integration.

Integration is defined in the dictionary as the act of combining parts into an integral whole. It is only necessary if you are dealing with separate parts and wish to combine them in some organized way. Systems are, by their very nature, combinations of parts that together perform some function that no subset of the parts could achieve. But, it would be possible for a single human being to conceive, design, manufacture, and assemble a small system. Would integration take place in this case? Certainly the individual would have to integrate the consequences of a broad range of ideas within his/her own mind, but would not have to integrate the work of two or more people or organizations.

We tend to think of the need for specialization only as a negative. Would we not be better off if one human mind could attack a problem and avoid the chaotic environment sometimes caused by people trying to work together? Well, it simply is not so. However painful your memories of past programs may be, there is a positive benefit hidden in the specialization and integration scenario. Why is this? There really should be little doubt that if a single individual was in competition with a team of specialists who were well led, the system created by the team would be superior. Even if one person could master all knowledge, that one person's mind would function in a particular fashion constrained by his or her experiences in life. Given that no two people are the same, the team must collectively have command of a greater body of knowledge and experience. What is engineering but the translation of today's technical and scientific knowledge into designs for useful things. More knowledge and broader experience well used cannot help but fuel more possibilities from which to select an ideal solution to a particular engineering problem. Some of these ideas will be better than others.

We find it necessary to specialize in a competitive situation because our limitations prevent any one person from understanding and applying everything known by man or generalists from efficiently taking advantage of knowledge in a timely and competitive fashion. The added benefit is that we are forced to apply more than one person's mind to a problem and gain the advantage of expanded experience. To realize benefit, however, we must be able to integrate the work of these teams of people.

2.2 The fundamental integration mechanism

The specialization answer to complexity in excess of an individual's capacity for knowledge requires many people working together to solve a common problem which none of them can solve independently. Obviously, human communication, therefore, plays a vital role in system integration. Integration takes place in the minds of the people working on a system development effort. It occurs, for example, when one specialist understands the need for design features required by another specialist which are in conflict with features he/she has identified and together they reach an understanding that satisfies both of their requirements adequately.

Somehow one of the specialists has to learn about the conflict and communicate it effectively to the other. That happens by communication through visual or verbal communications in a neutral and common language, shared by all of the specialists, followed by the application of reason, analysis, and synthesis on the part of the other specialist. The opposite communication path is exercised also so that both understand the other's perspective. There follows a conversation where both try to understand the real constraints from the other's perspective in search of common ground and the true extent of the conflict. Integration, despite the availability of powerful computer tools and elaborate checks and balances, is largely an interpersonal relations problem and solution.

Jointly, the two specialist are able to find common ground about two different requirements that neither could have mastered independently. This is the strength of the systems approach involving decomposition of big problems into smaller ones on which specialized persons may work effectively together. This process of working together requires effective communication of ideas and information. The greatest opportunity for breakdown in integration occurs at the interfaces between team or individual responsibilities. These are the critical interfaces alluded to earlier. Unfortunately many people tend to withdraw from these interfaces instead of plunging across to understand the other persons problems. A good rule of thumb in integration work is to encourage or require engineers to understand all of their product interfaces from the perspective of those responsible for the other interface terminal as well as their own. We take this matter up again in Chapter 12.

The whole teamwork notion that the integrated product development scenario promotes revolves around human communication because it is through communication that we understand together how to solve our common problems on teams. There is no grand mystery here and nothing very complicated. Specialized humans must communicate with each other to form the equivalent of one all-knowing person.

Implementation of integrated product development is, therefore, a study in how to improve communication between people. The most important step, aside from a common language, is physical collocation of the people working on a team. Research has shown that the amount of communication between people is clearly inversely related to the distance between them in the work environment. Meetings, good telephone and electronic mail system, and networked computer information systems offer other useful integration stimulants, but do not fully replace the need for physical collocation. In Chapter 5 we will discuss an alternative to physical collocation that is rapidly becomming possible, but for now let us accept that physical collocation is the best way to encourage communication.

We also need a degree of order for integration to be successful so as to ensure that necessary conversations do take place. To facilitate this we could introduce a very rigid structure with forms and checklists and rules, as some have done. It is not uncommon for system engineers to fall into the trap of over organization in this area to the extreme even of stifling creativity. Our goal must be to establish a degree of order that maximizes the ability to communicate while minimizing the need to communicate and to do so with the least negative impact on the creativity of our people.

2.3 The Dark Ages of system engineering

The system engineering community must approach facilitation of both communication and team structuring with some care. If we gain benefit from the diversity of our specialized people, could we not defeat ourselves by being too successful in organizing and standardizing our process? That is a real fear. It is entirely possible that we could so rigidly standardize our process that there is no room left for creative thinking. Some system engineers feel that during the 70s and 80s the Air Force Systems Command, responsible for development of new Air Force systems, dangerously approached this condition with its Air Force Regulation (AFR) 375 series of procedures.

The procedures in 375 called for otherwise intelligent engineers to fill out what seemed like an endless array of forms to document the results of the engineering effort. Some companies actually had one team of engineers working on the design problem and another team working on the forms so as not to contaminate the creative minds of the "real engineers" with what some considered busy paper work. Many surviving system engineers from this period feel that the 375 series poisoned the well for years against the practice of effective system engineering. People in management came to feel that the system engineering process, they associated with those damnable dehumanizing forms, did not contribute to the product design process.

It is during this same period, and certainly not solely traceable to the 375 series, that the system engineering profession lost its way, overcome by the ease with which we can more finely specialize and the difficulty with which we improve our ability to integrate the work of many people. Stovepipe specialized disciplines developed with dedicated support within Government customer ranks as well as in industry. Standards were prepared that defined in great detail how each stovepipe specialty would be accomplished. Many engineering organizations evolved into transom engineering shops featuring serial work performance. Designs were prepared in one shop, thrown over the transom for independent assessment by a crowd of specialists and this cycle repeated until the exhaustion of all.

As an example of this process, while the Advanced Cruise Missile (ACM) was in development at General Dynamics Convair Division in 1983, a design engineer responsible for the design of a portion of the airframe repeatedly offered his design solution to the nuclear survivability and vulnerability engineer for review. Each time the design engineer was told that it would not satisfy his specialty requirements. After the third turndown, the design engineer asked the specialist, "How are we going to know when I get it right?" To which the specialist replied, "I will know it when I see it."

This was a particularly good example of transom engineering because one solution to nuclear survivability and vulnerability requirement, added mass, was in clear conflict with the weight requirements of the mass properties specialist. The Chief Engineer had earlier encouraged the mass properties lead engineer to organize an active program of weight control that even included prizes for the most weight removed. Our structures engineer in this story was trying hard to satisfy the weight requirements allotted to him resulting in conflicts with the nuclear survivability and vulnerability requirements that he didn't understand. The general solution in this particular case was for the nuclear survivability and vulnerability engineer to hold a class for the design and mass properties engineers to explain his requirements including some ways that engineers had solved these design problems in the past. Thus communications and teamwork overcame transom engineering and we will see these same powerful and very inexpensive tools brought to bear throughout this book for the same purpose.

2.4 *Order vs. creativity*

Now, we should not hang this whole guilt trip on the U.S. Air Force. Actually, the 375 series of procedures contained a lot of good ideas, many of which were perpetuated in a series of Intercontinental Ballistic Missile (ICBM) weapons system requirements analysis standards and later used by the author as a major component of the model for the companion book *System Requirements Analysis*. The 375 series attempted to capture information for which a computer was really needed but we humans had not yet figured out how to take full advantage of a computer in association with this work. At the time this book was being written, the computers and the humans who can program them were ready to help solve this problem of order versus creativity. In the early 1960s when the 375 series was published, the computer apparatus was not available. Early attempts to mate the process to the computer were also flawed requiring the people to deal with precisely the same forms on the computer screen that were used in the paper implementation.

Computer applications are beginning to evolve that apply the computer's strengths to eliminate human drudgery while helping to identify the work where human creativity and thought can most profitably be brought to bear. The product RDD-100 created by Ascent Logic, Spec Writer by PRC, and other tools in the works by other companies offer graphical user interfaces (MAC and Windows desktop screens) that expose the human to useful fragments of the whole task and from which you can draw integrated views of the information contributed by a team of experts using workgroup computing. These tools are just scratching the surface compared to what we will be using in a few years to automate the system engineering process but they are an absolutely necessary step in the evolutionary process. They help to expose new directions that can be exploited and without them we would be unable to see new opportunities. In Chapter 5, we will reopen this discussion.

In our zeal to implement sound system engineering processes, we must guard against upsetting the balance between order and creativity in our organization. In hammering out an orderly process, we must not loose sight of the benefits to be derived from diversity, dispute, intuition, and creativity. When confronted with the thought that the fledging National Council on Systems Engineering (NCOSE) was unable to get anything done, the first President of NCOSE, Dr. Jerry Lake of Defense Systems Management College, responded that there was value in diversity. As much as we may be infuriated by the workings of Congress, the balancing measures established by our founding fathers have protected us against a lot of bad ideas. And so it is in system engineering.

While this book encourages close teamwork and problem solving by consensus, the process of arriving at a conclusion may not, and need not, be completely blissful. This process may be characterized by heated debate, wild excursions of the human mind, and all of the potential interpersonal relations problems that define human affairs. After all, we must work with the normal slice of humanity our company is fortunate to have in its work force. To give this force the best possible chance at success, we must be conscious of a need for balance between order and creativity. Either extreme is detrimental to the evolution of a sound system development capability.

We must establish sufficient order and discipline that everyone working on a program has a common framework for their work such that the information product of each individual fits into a pattern supporting one and only one configuration (parallel development of multiple alternatives in early program phases excepted). We must respect the schedule agreed upon with our customer and avoid analysis paralysis. But, we must also allow design engineers the freedom to conceive wild and crazy ideas consistent with the requirements, some of which might actually work.

Programs need to place the balance point far over toward creativity in early program phases and far over toward order and control subsequent to a design freeze during what DoD calls the Engineering & Manufacturing Development phase. The author likes to picture this as a mixer facet with hot and cold water valves. Early in a development activity you should have the cold water (order) on only slightly and the hot water (creativity) on in a mighty torrent. As the program progresses, these facets must be readjusted to protect the integrity of the evolving baseline until after the Critical Design Review (CDR), or commercial equivalent, the creative juices for change are tightly under control and pure cold water flows. At the same time, the earlier appeal to creativity should have previously led to the best possible cross-functional solution that will not later require changes.

2.5 *Mathematical chaos as an alternative*

Rather than being two extremes on a linear scale, it appears to the author that order and creativity are arcs on a circle where, if you go far enough in the creativity direction, you descend into the true opposite of order, chaos. But we are finding now that there is a beautiful order in chaos.

The author has long speculated, from a perspective of imperfect knowledge, on the potential for derivation of a system development model from mathematical chaos that might be superior to structured, top-down development. Chaos helps to explain evolutionary processes that have done a pretty fair job of designing us and our natural world. The question is, "Can this same process be applied consciously by us to develop man-made systems?"

Right now the answer seems to be that we do not have the time and money to spare for mathematical chaos to work its magic. Chaos appears to require a lot of trial and error which would translate to time and money generally not available in sufficient supply to encourage anything but an organized direct approach to a rational solution. It can be argued that chaos is actually at work today in a macro sense shaping our systems of systems. Weak system solutions to problems are replaced eventually by new systems and subjected to the trial and error evaluation of the real world. This process keeps repeating and when it works out well in a particular field we call it progress.

But, could the principles of mathematical chaos be applied with more immediate effect at the micro level of system development in our day-to-day decision logic, trade study approaches, and the pattern of human interactions now referred to as integrated product development? The author does not know the answer to this question. What is apparent though is that for the time being, the structured development approach is the best machinery we have to evolve what is needed to satisfy our needs. And system integration is an inseparable part of that process because of our need to decompose large problems into sets of small ones worked on by teams of specialists.

chapter three

Organizational structure

3.1 Updating matrix management

In order to discuss integration, we must make some assumptions about the organization through which we will deal. There are three principal organizational structures common in industry: (1) projectized, (2) functional, and (3) matrix. In the projectized organization, the personnel are assigned to each program which has hiring and termination decision-making authority. In the functional organization, all personnel are assigned to departments that specialize in particular kinds of work such as: engineering, production, quality assurance, etc. The matrix organization approach imperfectly attempts to marry the positive aspects of each of these approaches and avoid the negative aspects.

The author supports the matrix management arrangement for large companies with multiple programs. The matrix is characterized by: (1) functional departments led by a supervisory hierarchy, possibly including Chiefs, Managers, Directors, and Vice Presidents, which provide qualified personnel, tools, and standard procedures to (2) project organizations responsible for blending these resources into a set of effective product-oriented cross-functional teams called integrated Product Development Teams (PDT) and managing these teams to achieve program success measured in customer terms.

The functional departments provide a pool of qualified specialists trained to apply department-approved standard best practices and tools in the development and production of products appropriate to the company's product line and customer base. The functional departments are charged with the responsibility to continuously improve the company's capability through small improvements in training, tools, and procedures based on lessons learned from program experiences and continuing study of available technology, tools, methods, company needs, and the capabilities of competitors. Company programs are internal customers of the functional departments.

Program PDTs are organized about the product architecture reflected in the program Work Breakdown Structure (WBS) and/or product architecture diagram. The teams are selected and formed by program management. They are led by persons selected by the teams from the personnel assigned, with the approval of program management. Once identified, the team leader is responsible for: (1) molding assigned personnel into an effective team, (2) concurrently developing the assigned product requirements followed by (3) concurrently developing a responsive product design, test, manufacturing (tooling, material, facilitization, and production), operations and logistics, and quality concepts.

PDT personnel must first focus on team building matters in order to form an effective cross-functional group. Next, the team must focus on product and process requirements analysis. The product of this work should be documented and approved by program management. The team must, concurrently with requirements work, develop alternative concepts responsive to the requirements as a way of validating and demonstrating understanding of the maturing requirements. Where there is no clear single solution, the team should trade off the relative merits of alternative concepts. The team must be very careful not to influence the requirements identification too strongly by concept work but should take advantage of valid concept-driven requirements identified. This requires a delicate balance that should be assertively monitored by a systems PDT called here the Program Integration Team (PIT) and by program management at internal reviews. The requirements must also drive manufacturing, quality, logistics, material, test, and operational design work as well as product design.

Only after approval of the requirements and complying concepts at an internal design review, and at appropriate customer and/or Independent Verification & Validation (IV&V) reviews, if required, teams should be authorized to proceed with concurrent product detailed design work and test, manufacturing, tooling, material, operations, logistics, and quality process design work. The team must implement all of the assigned tasks within budget and schedule constraints producing documentation that clearly defines what must be produced and how it shall be produced, tested, verified, and used to comply with the pre-defined requirements.

Functional management staffs programs with qualified personnel appropriate to the tasks identified on the project. Personnel must be assigned to programs with a reasonable degree of longevity because PDTs require personnel assignment stability to be effective. The reason for this is that the principal work of teams is communications and communications networks are fractured by personnel turnovers and these fractures require time to heal.

Day-to-day leadership of personnel assigned to programs should be through the PDTs and their leaders rather than the functional departments. Personnel assigned to one team throughout a complete quarterly personnel evaluation cycle should receive an anonymous evaluation from their fellow team members coordinated by the team leader. Personnel assigned to more than one team in one evaluation period can receive evaluations from all teams within which they served. Each functional Chief should review all of the program team quarterly evaluations for persons from his/her department and integrate this data into department ranking and rating lists used as a basis for all administrative actions (training needs, compensation adjustments, promotion/status quo/setback decisions, and program assignment considerations).

Functional department supervisory personnel, and senior working personnel under their guidance, should monitor the performance of the personnel assigned to programs from their department and provide coaching and on-the-job training where warranted. Functional management also should provide programs with a source of project red team personnel used by the Program Manager to review the quality of program performance in the development of key products. Functional department Chiefs can then use this experience as a source of feedback on improvement needs appropriate to current standard procedures, tools, and training programs.

Functional Chiefs, Managers, and Directors should be encouraged to follow the performance of personnel from their department on programs, but should be forbidden to provide program work direction for those personnel. That is, functional management may provide help and advice for their program personnel in how to do their program tasks but should not direct them in terms of what tasks to do nor when to do them. Only the program oriented PDT leaders and lower tier supervision, if any, should be allowed to direct the work of team members

Functional management personnel, in this environment, are rewarded based on the aggregate performance of their personnel on programs (in terms of CDRL submittals, major review results, budget and schedule performance, and noteworthy personal efforts recognized by project management) and the condition and status of their department metrics, which may include depth, breadth, and quality of standard procedure coverage for the department charter, toolbox excellence, and personnel training program effectiveness.

Program budgets are assigned to PDTs by Program Finance. The PDT leaders are not only responsible for the technical product development tasks defined in an Integrated Master Plan and Integrated Master Schedule, but team budget and schedule. PDT leaders report to the Program Manager. Teams are generally, but not necessarily, led by engineering personnel during the early project phases (when the principal problem is product concept development) and later by production personnel (when the principal problem is factory oriented). Some companies have solved this leadership problem with engineering led development teams through first article inspection and factory oriented production teams thereafter.

3.2 *A model program organizational structure*

In our model of the perfect world, each project includes two or more PDTs and one Program Integration Team (PIT). Some people prefer the name system engineering and integration (SEI) team in place of PIT and these readers can insert that term or any other in place of PIT. The PIT should be led by a deputy Program Manager drawn from available senior personnel. During the early phases, this person should be someone with extensive Engineering experience. In later program phases, this position should be reassigned to someone with extensive production experience. The senior Engineering person assigned to the PIT will be called the Program Chief

Engineer in this book and he/she is responsible for monitoring all engineering work on the program and coordinating changes in engineering personnel team assignments through the PIT and PDT Leaders. The Chief Engineer is the principal product technical decision-maker for a program. Other major functional departments will have persons assigned to the PIT with similar responsibilities. The Chief Engineer is the engineering leader of the PIT and could be the PIT Leader as well in early program phases.

The PIT is responsible for technical direction and integration of the work products of the PDTs toward development of a complete product. This includes: (1) performance of initial system analysis and development of program level documentation (such as the system specification and other high level specifications, program plans, system architecture and interface block diagrams and dictionaries, and system level analyses); (2) mapping of the evolving system architecture to PDTs and formation and staffing of the teams; (3) review and approval of PDT requirements documentation; (4) granting authorization for PDTs to begin design work based on an approved set of requirements, concepts, schedules, and budget; (5) monitoring the development of interfaces between team items; and (6) development of interfaces between the complete product and external elements (system environment and associate contractor items).

The PIT, like the PDTs, reports to the Program Manager as illustrated in Figure 3-1. A project Business Team includes all of the project level administration functions such as scheduling, finance, configuration management, data management, personnel, program procedures, project level meeting management and facilitation, action item management, program reference document libraries, information systems services, and the program calendar of events.

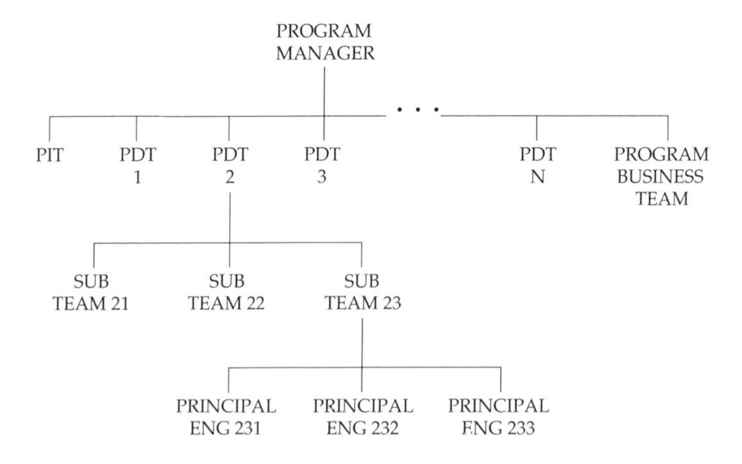

Figure 3-1 Program team structure.

PDTs organized at the system level may have to further decompose items for which they are responsible in order to reduce the problem space to workable proportions. In this event, the system level PDTs (those illustrated on the first tier of Figure 3-1 which report to the Program Manager) will create their own sub-teams, like PDT 2 in Figure 3-1, and acquire any additional personnel needed to accomplish sub-team tasks. Sub-teams may be fully staffed for independent work or rely heavily on the parent level team for specialists. Sub-team leaders are referred to either as PDT Leaders or Item Principal Engineers as a function of the magnitude of the task. At the lower system levels, such as an on-board computer or valve assembly, the person assigned development responsibility is referred to as a Principal Engineer (as under sub-team 23 in Figure 3-1). The Principal Engineers draw upon the specialty engineers assigned to the parent PDT (or sub-team) for related specialty work. PDTs are responsible for integration of the work of all sub-teams created under their authority. The PIT has this same responsibility over all PDTs at the system level.

During the early development phases up through Functional Configuration Audit (FCA), where proof is offered that the design solution has satisfied the pre-defined requirements, the Chief Engineer should normally lead the PIT and be the principal product system level technical decision-maker subject to management by exception by the Program Manager. The PIT is the internal reviewing and approving authority for the system PDTs. PIT must review each system team's requirements and concept sets before the team shall be allowed to proceed to concurrent design.

Each PDT Leader is the principal technical decision-maker for the PDT and should review and approve subordinate team requirements, concepts, and designs. Each PDT is responsible for development of all of its internal interfaces and jointly responsible, with the team responsible for the opposing interface terminal, for all of their external interface. This responsibility extends from PIT at the system level down through PDTs, sub-teams, and principal engineers as a function of the scope of their responsibility.

Phantom PDTs can be formed for items to be developed by associate contractors and work accomplished with this team through what is commonly called an Interface Control Working Group (ICWG). The senior member of this team will be selected for their interface development and interpersonal skills. On small programs the ICWG may be a subordinate task of the PIT.

Teams or sub-teams may also be formed around major supplier items such as a rocket engine, guidance set, or other high-dollar or schedule-critical item. Leaders of these major supplier items may report to the Program Manager like other system level teams or be organized as sub teams to a propulsion or guidance system level team. The Team Leader of a supplier team might be selected from procurement. These teams might be treated like principal engineer items as a function of the dollar amount, development complexity, schedule criticality, or program policy.

3.3 Resistance to PDT

There are many reasons why the organizational structure described here is very difficult to put in place in a company currently organized to perform autonomously within their functional organizations. These reasons include: (1) it disturbs existing organizational power relationships, (2) a conclusion that it is in conflict with the Department of Defense (DoD) cost/schedule control system (C/SCS) criteria defined in DoD Instruction 5000.2, Section B, Attachment 1 and the work breakdown structure concepts defined in MIL-STD-881B, (3) PDT creates staffing problems for the functional organization, and (4) PDT interferes with an efficient personnel evaluation process. Let us look at each of these reasons.

3.3.1 Human resistance

Many of those currently in functional management roles will perceive this prescription for distributed power as a personal, professional, and career threat. If they remain in functional management, they will see their relative power in the organization decline as it must to implement concurrent development. This will commonly result in resistance from those now in functional management roles. This resistance is inevitable. There is no best way to make the transition other than to announce it as far in advance as possible and openly discuss opportunities for those currently in power under the new system. It is, of course, entirely possible that at least some of those in functional management should not be allowed to survive the change.

The long range view of this shift, which many will not be open to see, is that a shift to the kind of organizational structure described above will attract a different collection of people to functional management positions. This position should focus on providing programs with qualified people trained in the standard techniques defined for a function and skilled with the standard tools supportive of those methods. Personnel selection and training, continuous process improvement, and tool building are the principal roles of this new kind of functional manager.

Many of those now in functional management positions will not wish to continue working in the new environment. Some will prefer to migrate to program centers of power (program office, PDT leadership, or program functional leadership roles). This is not a disaster, though some may perceive it to be; it is simply re-balancing of the power centers. Those for whom power is a motivator will migrate to it wherever it resides.

Those who have worked in a matrix management structure for many years have observed many shifts in the balance of power between the functional organization and program organizations. These are not uncommonly driven by the personalities of those in power rather than by some sound rationale for improvement in the company's capability. The change we are discussing here requires an enlightened management whose first priority is to satisfy the company's customers and needs for company profitability. They must have a willingness to make changes toward these ends despite a possible temporary setback in attainment of personal goals. This is a lot to ask of perfectly normal human beings populating management positions and it will not often be satisfied in the real world. On the other hand, the change may be a necessary prerequisite to the company's survival in a very competitive world and this can be a very good motivator.

Obviously, movement into PDT requires high level support and leadership in an organization. Anything short of total commitment at the top will prolong the transition or destroy best efforts from any courageous and selfless people at the lower level.

3.3.2 *C/SCS criteria conflict*

3.3.2.1 *What is the criteria?*

The Department of Defense has written a criteria, called the cost/schedule control system (C/SCS) criteria, for accounting and reporting systems that contractors must support in order to be considered for award of a government contract. The C/SCS criteria listed in DoD Instruction 5000.2, Part 11, Section B, Attachment 1 (formally in DoD 7000.2) includes five criterion: (1) organization, (2) planning and budgeting, (3) accounting, (4) analysis, and (5) revisions and access to data. Some companies dealing in the commercial marketplace may have similar two-dimensional accounting structures that may also create problems for implementation of the product team concept.

Some companies following the matrix management model have concluded that there is a fundamental conflict between the C/SCS criteria and integrated product development. The conclusion is that they must relate all program cost in a matrix between their functional organization and the contract work breakdown structure (WBS). Given that they wish to manage programs using integrated product development teams focused on WBS items, they cannot give the leaders of these teams control of the budget allocated to those items because they feel the criteria requires them to allocate it to cost account managers oriented toward their functional organization. Without budget control, the PDT structure is doomed to failure. In Chapter 11 we will discuss how to structure a WBS. For now, if you are not familiar with this entity, simply accept that it is a hierarchical arrangement of all program cost elements.

If there is a conflict between these criteria and PDT, it resides in criteria 1 or 3 as a function of how we interpret the phrases "functional organizational elements", "organizational structure", and "organizational elements". None of these terms are specifically defined in DoD I 5000.2, Part 11, Section B, Attachment 2. Definitions for the terms "performing organization" and "responsible organization" do not refer to either the functional or program structures of a matrix organization only to a ". . . defined unit within the contractor's organization structure . . ." DoD does not require a contractor to organize under any particular model (matrix, projectized, or functionally). So, the criteria must be implementable by companies following any of the three kinds of organizational structures.

The DoD I 5000.2 criteria elements that relate to direct costs and organizational structures are listed below. Note that the numbering and lettering of these steps is different between DoD I 5000.2 and 7000.2 but the content is the same except that the contract WBS acronym used in 7000.2 is spelled out in the newer version.

1. Organization
   b. Identify the internal organizational elements and the major subcontractors responsible for accomplishing the authorized work.
   c. Provide for the integration of the contractor's planning, scheduling, budgeting, work authorization and cost accumulation systems with each other, the contract work breakdown structure, and the organizational structure.
   e. Provide for the integration of the contract work breakdown structure with the contractor's functional organization structure in a manner that permits cost and schedule performance measurement for contract work breakdown structure and organizational elements.
3. Accounting
   c. Summarize direct costs from the cost accounts into the contractor's functional organizational elements without allocation of a single cost account to two or more organizational elements.

3.3.2.2 *Alternative approaches*

The premise is that there is a conflict between this criteria and the use of PDT, or what the DoD finance community calls a work team concept. Given that this is a true premise, there are several alternative arrangements we should consider in reaching a condition of compatibility between this criteria and PDT.

3.3.2.2.1 Status quo. Traditionally in a matrix organization, we allow the cost account manager (CAM), assigned from a functional perspective, to manage a functional department's budget and we hold him/her accountable for variances to a Program Manager responsible for one or more product-oriented budget segments, a role referred to in some companies (and in this book) as a WBS Manager. Total program cost is decomposed into intersections between functional departments and the WBS. Program Office WBS managers manage the WBS columns of the matrix and CAMs manage the organizational rows of the matrix. The PDT structure seldom overlays this matrix in total alignment with either matrix axis. PDT leaders are cut out of the budget management process and cannot possibly manage their team without budget control.

3.3.2.2.2 Change the criteria. The DoD finance community and many other customer organizations are wedded to the criteria till the end and maybe properly so. It was born out of contractor abuse and it will require decades of highly ethical behavior on the part of contractors to erase the stain of past excesses. Let us assume that we cannot change the criteria in any substantial way nor can we change it in any way in the near term.

3.3.2.2.3 Functional organization suppression. The criteria does not attempt to define what a functional organization is so we could suppress our functional organization in a matrix organization and establish a matrix between the WBS and PDTs on programs. Probably a suspicious government finance analyst would consider this a perversion of the criteria motivated by an attempt to avoid scrutiny in some way. Certification of our cost control system would be hard won. It would also make it difficult for functional management to get insight into their personnel needs since there would be no way for them to accumulate or project future demand for their department's services.

3.3.2.2.4 Three axis matrix. It appears we have three axes of interest in our data, PDT, WBS, and the functional organizations, and we are trying to manage with a two dimensional matrix. So why do we not deploy an accounting system that can handle three axes or a mapping of budgets three ways instead of the present two? While it does not appear that this would be out of step with the criteria (we would have an organizational structure to which we could map budget), it would likely be hard to certify the system because of its unusual nature. This would also require some new development for the accounting system.

A variation on this theme could assign pseudo department numbers to PDTs on programs, but other systems interfacing via department numbers may be disturbed. This alternative has a lot going for it actually. This approach can be implemented by simply assigning a unique department number to each of the PDTs on a program. As the program comes alive, we transfer personnel from their home functional departments to the PDT departments. Each PDT department manager can function as a CAM giving them the control we are seeking. The program office WBS Managers provide control from the WBS perspective.

3.3.2.2.5 Projectized organization. There is nothing in the criteria that requires us to organize in a matrix. Our company can be projectized, which would allow us to assign the organizational matrix axis to our program PDT structure. Such a company suffers from a difficulty in making continuous improvements because it has no one methodology — no central focus for its specialized functions. Some companies avoid this problem by splitting off each program into a separate business unit and allowing them all to develop their own identity and process.

3.3.2.2.6 Power to the WBS manager. It is not uncommon for the contract WBS to be under the authority of the finance community in government and contractor organizations. This results in customer finance people coloring technical system composition decisions by virtue of forcing a particular product organization before the engineering community has evolved the most cost-effective product organizational structure based on the need. As a result, the WBS may be at cross purposes to the product organizational structure about which the contractor would like to establish PDTs.

Customer and contractor finance computing systems are not sufficiently flexible that they can tolerate adjustment of the WBS to align with a technically superior structure, so the program moves forward with little interest in the WBS by the technical community. The WBS manager, therefore, can become a figurehead position in the program office divorced from direction of the technical effort focused only on cost and schedule concerns. Meanwhile, the PDT leader, focused on the technical aspects cannot control his/her team because he/she has no budget authority. No one person has the authority to manage integrated product development.

The right answer to this problem gives budget and schedule responsibility as well as technical responsibility to PDT leaders. The first step in assuring this capability is to place the responsibility for at least the product component of the WBS in the hands of the PIT. This is not to say that the finance community should be stripped of all voice in this matter. Rather, the decisions on WBS must be cross-functionally derived and not influenced totally by a finance position. Finance should be represented on the PDT and thus have a channel through which to communicate their valid concerns. This joint responsibility between contractor finance and the development team cannot be successful if the customer finance people stonewall and insist on a predetermined WBS.

In the process of making WBS decisions between alternatives, the PDT should weigh both financial and technical aspects of any particular choice but almost always choose the alternative that best enables PDT. This alternative will generally be characterized by minimizing or simplifying the cross organizational interface intensity between elements under the development management by different PDTs.

With this arrangement in place, we can then proceed to merge cost, schedule, and technical control of PDT into one person. Let us accept that we need to retain the functional organization axis of our matrix organization in our cost accounting matrix. This means that we must use the WBS axis to provide for PDT leadership and control. Instead of making functionally oriented CAMs responsible for budget management, we associate with this job only a responsibility to collect costs for functional personnel planning and staffing decision-making and financial reporting. Since the product WBS is structured in accordance with the product functional and physical structure, we now have alignment between WBS and the basis of assigning PDT responsibilities to the organization of the product. The WBS manager is the obvious person to take on the role of PDT Leader with cost, schedule, and technical leadership.

Under this arrangement, functional organizations cannot be held accountable for cost variances. Since the WBS manager is in complete control, the WBS manager must be accountable for variances. Functional managers and directors should not be unbraided for failure to satisfy program schedules and cost targets. This is a WBS manager responsibility, a program responsibility. Functional managers should be held accountable for providing qualified people, good tools, good procedures, all three of these being mutually consistent, and the continuous improvement of them based on program lessons learned. This results is a very lean functional department staff which is in step with current trends.

3.3.3 PDT-Stimulated personnel staffing problems

Some people hold that PDT creates staffing problems for the functional organization and this is true if in moving to PDT the functional organization is stripped of any knowledge of future budget availability. The functional organization needs this information in order to be able to ensure that the right number of people with the right kind of training and experience are available when the time arrives to use them on a program.

Some of the alternatives we considered above, frankly deprive the functional organization of the information it needs to satisfy future program needs. The last one considered and recommended, does preserve functional management access to this knowledge.

PDT can also result in PDT leaders contracting with the wrong functional organization for a particular kind of work. Some small companies may be able to handle the resultant volatility and it can even be a source of good restructuring ideas in a rough and tumble, but potentially effective fashion. Large companies will generally have difficulty with this and should have a clear definition of functional department charters with energetic enforcement of those charters by functional management when program management attempts to deviate.

The reason that these boundaries have to be jealously guarded is that you wish to hold the functional organization responsible for continuous process improvement concurrently with high standards of performance. A functional manager cannot be held accountable for maintaining a company's proficiency in a particular specialty if a different department is being contracted to do that work on programs. Suppression of deviations from the functional charter responsibilities is not necessarily, however, the best response in all case where this problem arises.

It is not uncommon that these occasions will be driven by a fundamental flaw in the way the company has assigned charters for specialties. So, when an incident occurs, functional and program management should first consider if there is some value in considering an alternative organization charter responsibility map. If there is not, then the functional perspective should almost always win out. At the same time, during periods when one or more disciplines are understaffed with respect to the demand for work, they may find it useful to establish temporary agreements with other departments to provide personnel to do their work. This same arrangement can be useful at other times as a means to create effective system engineers or simply to increase sensitivity to the needs and concerns of other disciplines while creating personnel with broader qualifications to improve flexibility in satisfying changing program needs.

3.3.4 Personnel evaluation problems

Now, we know that some people, Dr. Demming chief among them, are convinced that the devil himself/herself designed the personnel evaluation system used in much of industry. This system results in functional management ranking all department personnel based on the functional manager's perspective of relative worth. This ranking is then used as a basis for salary increases, promotion lists, and, to some extent, work assignments that can influence future evaluations by virtue of experience gained through assignments. It is not uncommon that the evaluation criteria is flawed and applied in an irrational or uninformed fashion besides. Commonly functional managers simply wish to get through this exercise as quickly and as easily as possible.

If you remove people from their functional organizations and physically collocate them with PDTs, you make it difficult for functional managers to observe their performance. If we believe the current PDT/TQM literature we accept that performance evaluation should be done by team members and not by functional management anyway. So, how do we provide for personnel evaluation?

If we preserve the functional axis, we must accept that functional management must be responsible for evaluation because they have the responsibility to provide programs with qualified people. However, functional management need not be the only source of input for evaluation data. If an employee were assigned to one and only one PDT for a whole evaluation period (typically a year), then the evaluation could conceivably be done by the PDT members in some fashion as Dr. Demming would suggest.

What happens when a person works fractionally on two or more teams over the evaluation period? It will not be uncommon for some specialists in traditionally low budget specialties to work like this. We simply need some mechanism for the accumulation, merging, and distribution to functional department chiefs of team-derived evaluation data. Also, it would likely be useful to increase the frequency of evaluation events to perhaps quarterly with this data merged in some way into annual figures for pay and promotion determination.

This is a valid concern for entry into a PDT work environment, but we should have little difficulty working up a computerized approach to acquiring the data from people assigned to teams, collecting and assessing that data, and providing it to functional managers in a fashion that encourages rational decision-making in rewarding personnel for performance. It will require a change from present methods that generally are arbitrary, counterproductive, and otherwise just terrible. Good riddance!

3.4 Model matrix for this book

Figure 3-2 illustrates the overall organizational structure we will assume in later chapters dealing with integration specifics. One of several program panes is expanded with program reporting in the vertical dimension and functional reporting in the horizontal dimension. The Program Managers report to the company or division executive as do the top people in the functional organizations on the left margin. These people taken together are called an integration executive to emphasize their principal function. A staff

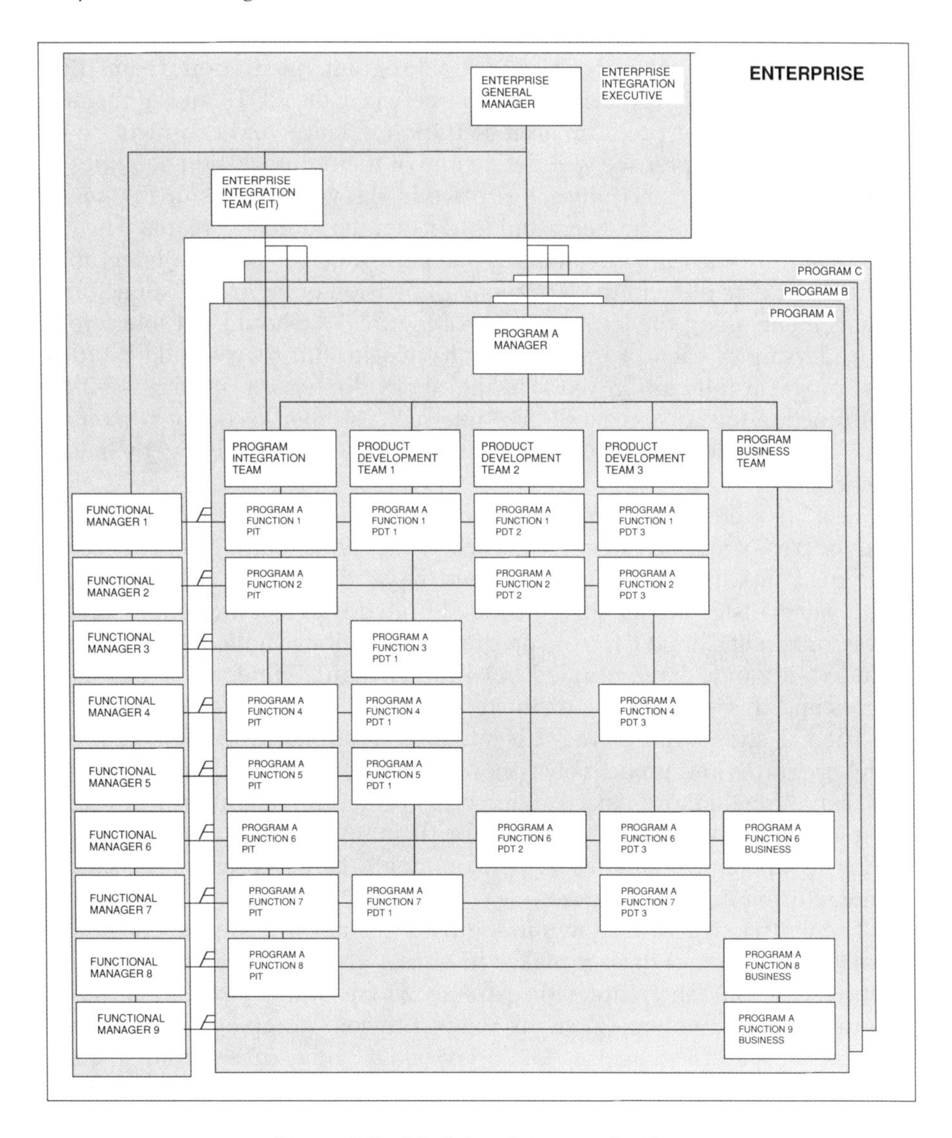

Figure 3-2 Model matrix organization.

Enterprise Integration Team (EIT) performs day-to-day enterprise resource integration through the program PITs under the oversight of the executive composed of senior management.

Each program has two or more integrated product development teams aligned with the product structure that is coordinated with the WBS. Each team is staffed by personnel derived by the program from the functional departments. In many companies, the functional Chiefs would control the tasking of the people staffing the intersections of the matrix. In our model, we will insist that the functional department Chiefs specifically not provide program task assignments or work direction. That must come through the program channel traceable to a WBS Manager serving as a PDT Leader.

Each program pane also includes a Program Integration Team (PIT) responsible for system level integration between the PDTs and a business team responsible for program administration, finance, and contracts.

In the next chapter we will define one of many integration activities as cross-functional. This activity takes place in the vertical axis of Figure 3-2 performed by system engineers and leaders on the indicated teams. The PIT members from each functional discipline perform cross product integration within their discipline in the horizontal dimension of Figure 3-2. Functional management, along the left boundary of Figure 3-2, should not interfere in the leadership of each program as we have said. But they should perform cross-program integration work in the depth dimension of Figure 3-2 for their function to stay abreast of their internal customer needs for personnel, tools, and practices, and to ensure all programs are applying their best practices well.

This book focuses on program integration and does not expand on the need for cross program integration work by functional management. It could be argued in Chapter 4 that we should have identified it as one of four components instead of the three included (functional, product, and process). In the context of Chapter 4, cross program integration is nulled. If it had been included, it would have resulted in 64 rather than 27 integration spaces to discuss and this would have distracted from the intended focus.

The EIT shown in Figure 3-2 is intended to monitor the resource needs of the programs and proactively seek out future potential conflicts and risks and then mitigate those risks through cross program integration work seeking a condition of the greatest good for the aggregate customer base. The principal function of this team will commonly be coordination of shared manufacturing and test facilities.

The matrix structure shown in Figure 3-2 is not intended to encourage matrix management where it makes no sense. The structure illustrated is a limiting case on the complexity extreme where one company has many programs each involving systems with sufficient complexity to warrant multiple development teams. Let us consider some other limiting cases creating an envelope within which every company will find their particular situation.

In the event the company in question only has one program in the house during a particular period, it only needs one program pane and the PIT can be merged with the EIT. If a company has multiple programs but each program is for a relatively simple element, they may need only one team on each program. In this case, the PIT is absorbed as a part of each team in the form perhaps of a single system engineer responsible for requirements definition and integration. The EIT is needed because we have multiple programs but the EIT will coordinate through the program system engineers performing the PIT function.

In the case of the ultimate simplicity of one simple program not warranting multiple teams, only one pane shows, the general manager and program manager become one, and there is one team composed of the functional managers and their personnel. The EIT is composed of the management staff who may have one person to implement their decisions. So, the author is not advocating the maximum matrix for all situations. Rather a flexible application of this structure is encouraged that matches the particular business situation.

A corporation may also find that it needs to add one more layer to this structure to integrate the work of several divisions. This same concept of structural flexibility can be extended to this higher level as well. As particular divisions become very successful in expanding their product and customer base, they may have to be partitioned into more efficient business units. As the business level retracts, it may be necessary to recombine some separate business units in search of a critical mass.

3.5 The virtual team

We have just painted a picture of necessity for the members of product development teams to be physically collocated in order for them to be able to take advantage of the resultant synergism. The reader should note that even if a team is physically collocated they can be badly led such that this desired synergism of group effort does not take place. The team could conceivably interact in a series of one-on-one discussions in which the other team members did not participate. So, physical collocation by itself does not a concurrent development success make. It also requires good leadership that encourages effective group communications and use of effective communications tools, about which we will say more in Chapter 5.

But, let us question the absolute necessity of physical collocation. Is there a way that a team can function comprised of people at least some of whom are physically or geographically separated by some distance to the extent that direct interaction is not possible without special communications aids? Yes, it is possible and some companies are already working at the fringes of what will become a very normal condition once the necessary interfacing wiring and fiber optics networks are extended to homes through the information superhighway that was in the planning stage in California and other places as this book was written.

On the day that this passage was written, it was possible for the author to interact with others over vast distances via telecomputing in one-on-one situations. But, we know we must empower the group dynamic, with groups larger than two, because of our narrow specialization forced by our limitations and expanding knowledge. In Chapter 5 we will return to this problem looking for solutions. For now let us assume that an effective team telecomputing capability is a near term reality.

Some companies that have manufacturing and design facilities in two or more countries are working effectively with integrated product development today empowered by good teleconferencing, videoconferencing, and telecomputing sharing of electronic data. But in these cases while the full team is composed of nodes of people at physically separated places, they are collected into nodes at companies. The leap that will soon be possible is to form teams of people working out of their homes. The difference between then and now is that few homes are wired into the kind the information grid with the kinds of circuits that can handle the bandwidth necessary for simultaneous near full motion video, voice, and computer data. These capabilities will be available soon.

Even today in late 1993, however, it would be possible to support a dispersed PDT with everything but good interpersonal video. Telecomputing from homes to central company computer networks by people on a conference telephone call is entirely possible. The one missing piece some would say is unnecessary, while others, including the author, would say is critical. That remaining piece is the visual integration of the group through full motion video.

Picture for a moment our PDT of the future. Our team members work at their home computers under flex time rules with an agreement that they must all be on their machines during the hours of 10 to 11:30 AM and 12:30 to 2:00 PM each work day based on the time zone of the principal work site. Throughout their work day all of these people will be tied into their program central database (see Chapter 5) and perform their assigned tasks adding to the central data store from their area of specialty.

At 10:00 AM (in the principal site time zone) each work day all team members connect to the conference bus and join a tele-video conference discussing team status and progress. As each person speaks, his or her picture comes into view in a corner of the screens of all team members. The screens all reflect the view that the speaker is addressing. All members are simultaneously hearing and seeing the same information stimulating new thoughts in the team members. These ideas are put into words and pictures and communicated to other team members. What results is exactly the same effect that occurs if these people had all been in one room at the principal site. The team members are virtually in that single meeting room and have become a virtual team.

The author first became aware of this possibility through discussions with a colleague Mr. Gene Northup, Executive Vice President of Leading Edge Engineering, Inc. based in San Diego, CA which was founded on the principles discussed in this section to provide companies with engineering services through remote computer communications. The concept was also covered in the February 8, 1993 issue of Business Week in an article titled *The Virtual Corporation* by John A. Byrne and in a book of the same title by W. H. Davidow and M. S. Malone published in 1993 by Harper-Collins.

Many companies are working now to introduce and perfect their application of the integrated product development approach with physical collocation thinking that it is some kind of terminal condition of excellence. The reality is that change will continue to flow over us in powerful surges with decreasing periodicity. The virtual team is a good example of that reality. While some are working toward PDT as the ultimate structure, others are working toward the next step of the virtual team and the flexibility it provides. In creating and maintaining our process, we must not fall into the trap of thinking in terms of discrete terminal events. We should think of our process improvement activities as a continuum upon which we have constructed a planned growth path that takes advantage in the near term of available communication, computer, management, and process technology while planning future possible paths based on research that we consciously seek out and experiment with.

An excellent way to stay abreast of new approaches is membership and active participation in a national society that shows an interest in this area. Some that come to mind are the National Council on Systems Engineering (NCOSE), American Institute of Astronautics and Aeronautics (AIAA), Institute of Electrical and Electronic Engineering (IEEE), American Electronics Association (AEA) and Society of Automotive Engineers (SAE).

chapter four

Integration components, spaces, and cells

4.1 Setting the stage for integration decomposition

It would be convenient if we could describe all of the facets of system integration in terms of a single entity. The author is convinced, after years of work, observation and study in this field, that many of us accept incorrectly that integration is one activity. Few of us are able to describe how that single activity is performed, however. Many people assign the term "system integration" a mystical quality. Whatever it is, it is the answer to every system engineering problem and seemingly just connecting the term in the same sentence with the problem is sufficient to define an approach to the concept in our proposals and conversations.

We will see in this chapter that the integration process can be decomposed into several parts and these parts explained in an uncomplicated way. System Integration then becomes the sum of those parts. We must, of course be watchful that we do not loose contact with important information about the whole when thinking in terms of the parts.

We begin with an acceptance that an engineering organization that must deal with multiple system development activities, should organize in a matrix structure, the subject of Chapter 3. The matrix has the advantage of focusing day-to-day work on specific program problems while providing a good environment within which to improve the organization's skills, methods, tools, and knowledge. We should also accept the good sense that we cannot beat the odds on specialization. Our engineering organization must select its personnel from the same pool as everyone else, humanity. We have seen that we are knowledge limited and we solve that problem through specialization.

Therefore, we will organize our personnel into functional specialties led by functional department Chiefs. These Chiefs will be responsible for providing all of our programs with qualified personnel, skilled in using a particular toolset and following the standard department procedures proven effective on past programs. We insist on standards, that are continuously improved, because we wish to take advantage of the practice-practice-practice template used by great athletes.

Our product will be organized into sub-elements that can be worked on by two or more Product Development Teams (PDT) and we will assign our personnel to these teams which will form the principal personnel supervisory structure. The work of all of the PDTs will be coordinated by a Program Integration Team (PIT).

We will organize all program work into process steps linked to product entities under the responsibility of one or more PDTs. Each process step will have a set of goals and a simple task description. We will map our PDT responsibilities to the processes and identify leaders for each process. All of the processes will be laid into our integrated schedule with clear start/stop dates and budgets which reflect back into the PDT structure. Each process will have clearly identified information and/or material product outputs that are needed in other processes as inputs.

4.2 Integration components

These assumptions and selections leave us with the problem of combining, or integrating, the work of many people in different functional disciplines, working on different product system components, in many different process steps over time. We define these three fundamental integration components as function, product, and process respectively. Be very careful that you understand that we have used the word function to mean functional organization and not product system function.

In each of these components, we have to concern ourselves with two fundamental kinds of integration: co-component and cross-component integration. In addition, we have to account for the possibility that one or two of the components is not involved. This means we have a three-valued situation: co, cross, and null values. With three variables, each with three values, it is obvious we are working with 27 (3 cubed) different integration possibilities. It is helpful to have a picture of this problem to better understand all of these possibilities.

Unfortunately, we can't easily illustrate the three-valued relationship for each of our three components, so our visual model will disregard the null possibility for each component. Figure 4-1 illustrates the three components as a three dimensional system. We can imagine that we can assign positions on the function axis to discrete organizations in our functional organization such as reliability, structural design, quality assurance, etc. Similarly, we can assign positions on the product axis to the elements of the system (avionics system, on-board computer, circuit card, transistor, etc.) in a hierarchical fashion, Finally, the third axis can have positions marked out corresponding to the processes on our program process diagram, such as identify item X requirements, design item X, and test item X.

For a given functional organization, all of the work that a specialty discipline does can be thought of as being within the plane passing perpendicular to the function axis at that function position. The task of integrating all of the work on that plane is called co-function integration, the integration of work accomplished by two or more specialists in that one functional discipline.

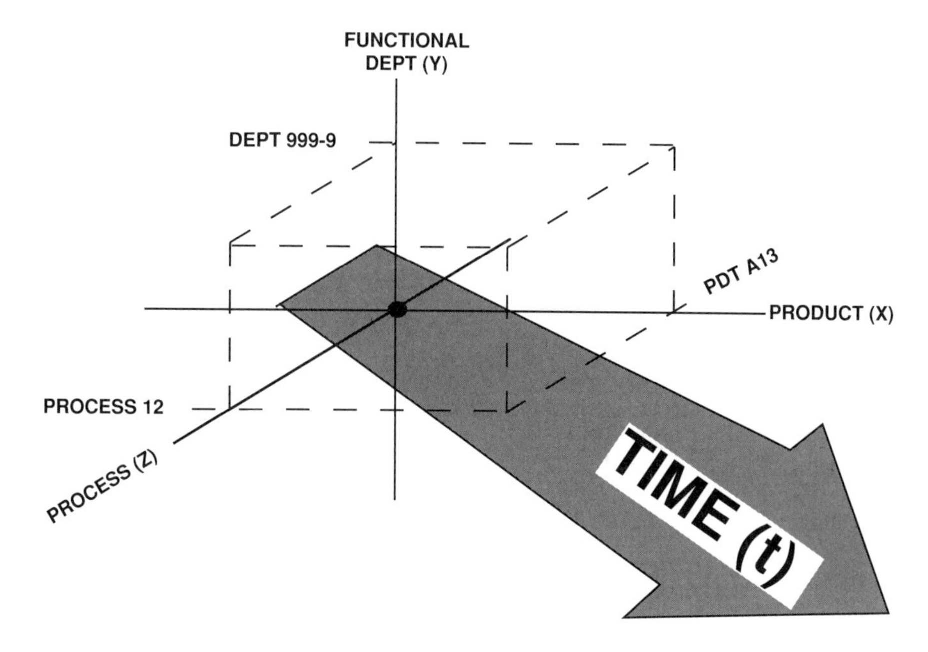

Figure 4-1 Integration components.

Similar planes can be constructed perpendicular to the other two axes at different points on those axes to capture all of the co-process and co-product integration work. The careful reader will observe that every point in the three-space, then, corresponds to some combination of co and cross component integration for the three components.

Let us now define the three integration components in terms of their possible values. Three combinations are nulls, meaning no integration for that component: Null-Function, Null-Process, and Null-Product integration. The other six possibilities are co and cross combinations with each of the three components. Let us take each of these six cases in turn assuming, in each case, that the other two components have a null value for the moment.

Co-function integration coordinates the work of two or more persons from the same functional department (specialty discipline) to ensure they are all using the same tools, techniques, and procedures in an appropriate fashion and that their results are consistent with other work on one or more programs or systems. This task is the responsibility of the senior program functional specialist for the project or the functional supervisor, in the case where all company projects are the target of the integration work.

Cross-function integration coordinates the work of persons from two or more functional disciplines in search of suboptimal design and specialty engineering solutions needing re balancing, mutual conflicts between specialty requirements or the corresponding design solutions, available unused margin to be repossessed and applied more effectively, and wayward interpretations of the project or product requirements that may lead to conflict. This is a program responsibility that may fall upon a PDT Leader, task principal, or person from the Program Integration Team (PIT) as a function of the relationship of the work to process or product and the integration level.

Co-process integration coordinates the work of two or more persons working in the same program process to ensure that they all are focusing on the same process needs within available budget and schedule constraints. This task is accomplished by the assigned task leader who is responsible to achieve the task goals on time and on budget.

Cross-process integration coordinates the work of two or more task teams working different processes. It seeks to ensure that the work of the two or more teams is mutually consistent and driven by the same program goals. This integration task is the responsibility of a program person

Co-product integration coordinates the work of two or more persons developing the design solution for a particular product item. This may be to develop an integrated set of product item requirements, evolve an optimum synthesis of those requirements in terms of a design concept, ensure that all of the cooperating specialty views are respected in the design solution, or integrate the test, analysis, and design work associated with that product item. Where an item requires a special test article, such as a flow bench for a fluid system, this work should also ensure that the test article properly reflects the same requirements and design embodied in the product item.

Cross-product integration is the commonly conceived integration component that most people would first think of in response to the word integration. It focuses primarily on integration of interfaces between product items to ensure that all interface terminals and media are compatible both physically and functionally. This work could be as simple also as ensuring that all items are painted the correct color, have satisfied a particular specialty requirement, as in being maintainable defined by an ability to remove and replace in 10 minutes. It is difficult to talk about this form of integration without linking other integration components.

4.3 Integration spaces

Integration work seldom falls into one of these pure cases. Commonly we have to deal with more complicated situations involving combinations of two or three components with mixes of cross, co, and null values. Figure 4-2 offers four particular examples of these possible combinations, or spaces, for further discussion. Once again, we disregard the null value case to enable simple graphical portrayal in three dimensions on two dimensional paper.

In Table 4-1 we use a simple tertiary counting scheme to ensure we have not omitted any combinations of the three variables. The reader should understand that there can be no other forms of integration work than those listed in Table 4-1 given that we have included every appropriate integration component. It would be possible to add one or more components to our list in addition to function, process, and product. The effect would be to multiply

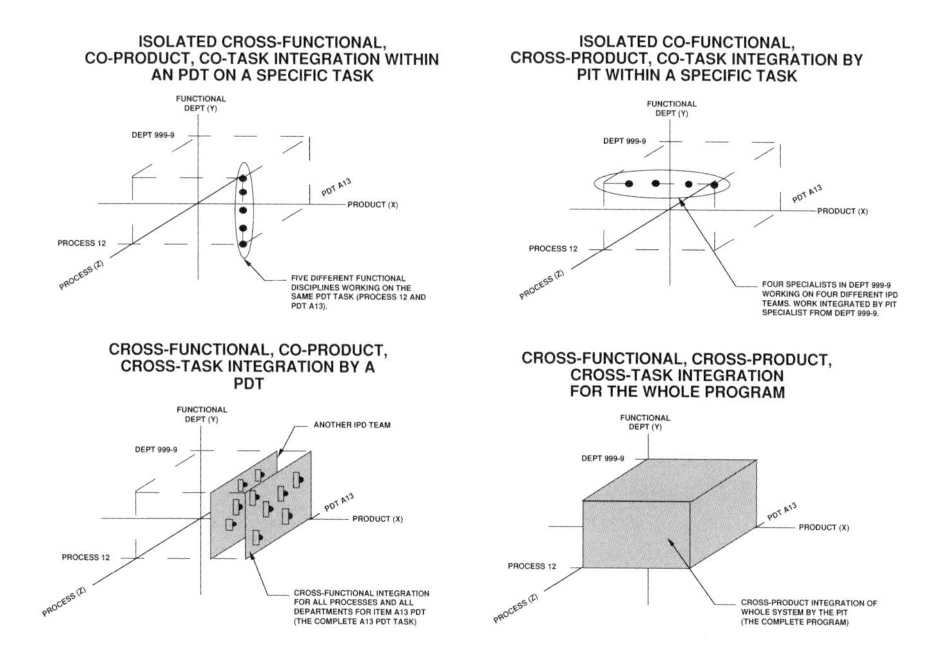

Figure 4-2 Integration space examples.

the number of different integration spaces. The number of integration spaces (S) is, of course, predictable as follows:

$$S = C^n$$

where C is the number of integration components and n is the number of values for each variable.

If, for example, we concluded that there should be five components instead of the three covered in this book each having three values (co, cross, and null), we would have 5^3=125 integration spaces. At the end of the last chapter we speculated about recognizing program as an additional component and noted that it, in combination with the three covered here, would require 64 integration spaces (4^3). These 64 spaces would include co, cross, and null program integration as well as all of the new combinations with the other components. In the context of the new ideas you have just been exposed to, the author elected to set this component at a fixed value of co-program and focus on a single program. As you can imagine the identification of additional components could get out of hand very rapidly making the description of integration more complex than the work itself. Three components and three values appeared to the author to be a good compromise between completeness and understandability for the purposes of this book.

We will use the INTEGRATION TYPE NAME in Table 4-1 throughout the book to refer to these combinations of integration components, referred to as integration spaces. The integration space ID numbers are used in a

Table 4-1 Integration space identification.

ID	FUNCTION	PROCESS	PRODUCT	INTEGRATION TYPE NAME
0	NULL	NULL	NULL	INDIVIDUAL EFFORT
1	NULL	NULL	CO	ISOLATED CO-PRODUCT
2	NULL	NULL	CROSS	ISOLATED CROSS-PRODUCT
3	NULL	CO	NULL	ISOLATED CO-PROCESS
4	NULL	CO	CO	CO-PROCESS & PRODUCT
5	NULL	CO	CROSS	CO-PROCESS/CROSS-PRODUCT
6	NULL	CROSS	NULL	ISOLATED CROSS-PROCESS
7	NULL	CROSS	CO	CROSS-PROCESS/CO-PRODUCT
8	NULL	CROSS	CROSS	CROSS-PROCESS & PRODUCT
9	CO	NULL	NULL	ISOLATED CO-FUNCTION
10	CO	NULL	CO	CO-FUNCTION & PRODUCT
11	CO	NULL	CROSS	CO-FUNCTION/CROSS-PRODUCT
12	CO	CO	NULL	CO-FUNCTION & PROCESS
13	CO	CO	CO	ALL CO
14	CO	CO	CROSS	FOURTEEN
15	CO	CROSS	NULL	CO-FUNCTION/CROSS-PROCESS
16	CO	CROSS	CO	SIXTEEN
17	CO	CROSS	CROSS	SEVENTEEN
18	CROSS	NULL	NULL	ISOLATED CROSS-FUNCTION
19	CROSS	NULL	CO	CROSS-FUNCTION/CO-PRODUCT
20	CROSS	NULL	CROSS	CROSS-FUNCTION & PRODUCT
21	CROSS	CO	NULL	CROSS-FUNCTION/CO-PROCESS
22	CROSS	CO	CO	TWENTY TWO
23	CROSS	CO	CROSS	TWENTY THREE
24	CROSS	CROSS	NULL	CROSS-FUNCTION & PROCESS
25	CROSS	CROSS	CO	TWENTY FIVE
26	CROSS	CROSS	CROSS	ALL CROSS

database designed by the author to experiment with system engineering information relationships and are included here only to illustrate the simple tertiary counting scheme for ensuring completeness. Each integration space involves a merger of a unique grouping of three particular values for the components and represents one integration mode useful in one or more particular situations.

Of the 27 integration spaces, we can dispense with INDIVIDUAL EFFORT integration immediately. We must assume that a single individual is fully capable in their field of specialized knowledge and able to effectively apply this knowledge to a specific product item and in a particular single process. This is why we specialized, after all, to create a human task that is within the power of a normal, single, specialized individual to master. We assume that each specialist can carry on an internalized conversation with themselves and use the power of their specialized discipline to solve the small problems we have tried to frame in our decomposition efforts.

Similarly, the ALL CO integration space is usually not very interesting to a system engineer since it involves people from the same functional department performing work in the same process step, for the same system element. True, on a very large program, we could imagine how this could become quite a challenge. For example, we might assign seven technical writers to the job of writing a single technical manual for a single item of equipment. Integrating the work of these seven writers would be a case of ALL CO integration.

One other fairly simple integration space to explain, though the most complex of them all in practice, is ALL CROSS integration involving cross everything. When a member of the PIT integrates the work of several members of two PDTs developing the design of two different product elements and the work of a facilities engineer responsible for the factory that will assemble these two elements and a tooling designer responsible for the manufacturing equipment that will hold the elements during mating, we have an example of this kind of integration. The reader will be able to imagine several other cases of this integration space.

We have given examples now for nine of the 27 integration spaces: including the six isolated integration cases, INDIVIDUAL EFFORT (or ALL NULL), ALL CO, and ALL CROSS. This leaves 18 remaining to be explained with at least one example. Some of these 18, you can see from Table 4-1, are very difficult to name so we will simply use the ID number for a name. Examples for most of these other 18 spaces will appear in the remaining chapters of the book.

4.4 *Integration cells*

In almost any given system development work situation, we will find it necessary for some combination of the 27 integration spaces to be applied. Very little system development work can be accomplished in total autonomy today. Returning to Chapter 1, we recall that this phenomenon appears because we have had to specialize very finely to master enough of the available knowledge base to be competitive. The number of these combinations is finite but quite large for a large development program.

We have many options in grouping all program work into unique combinations of integration spaces but the best way to do this is probably driven by the program tasks that would appear on a program task network. Each program task has associated with it some set of functional disciplines performing work on some particular combination of product elements. Many of these tasks, possibly most, will require some form of integration in the context of some combination of the 27 integration spaces defined above. Let us call any one of these combinations of task, product, and functional organization an integration cell.

The more finely we divide the overall program into tasks, the more unique integration cells we will have. The more finely we assign PDTs to

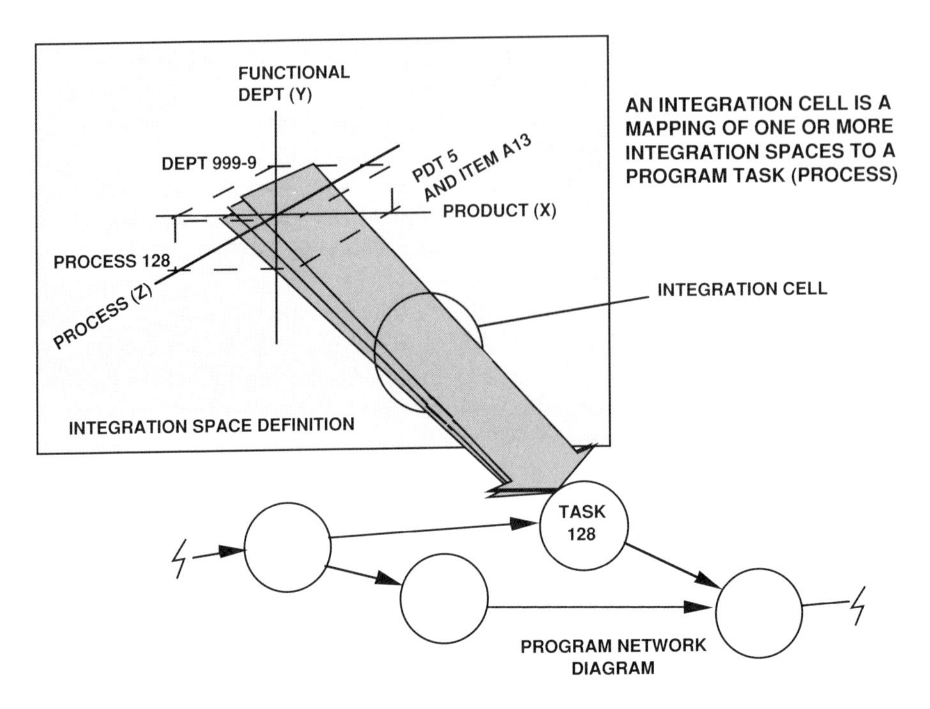

Figure 4-3 System development world line.

develop the system, the larger the number of integration cells. The more functional disciplines that must be assigned to the program, the more integration cells that will be necessary. At the same time, the larger the number in any of these integration components, the more simply we can describe each integration cell because they will consist of a less complex combination of integration spaces.

4.5 *Program world line*

So, the integration process is more complex than we might have first imagined, composed of a finer structure than we might have thought. But the complexity does not stop here. We must not only apply each of these integration spaces well within the context of the integration cells defined for the program, but we must apply them in a pattern coordinated with the program schedule. Figure 4-3 illustrates this need by placing the integration spaces coordinate system on a program world line. Throughout program passage on its world line (network or schedule), appropriate specialists must apply the appropriate integration spaces to the planned integration cells in accordance with an integrated plan. To the extent that we can reduce aggregate program work by early identification and resolution of inconsistencies between product and process elements, we encourage success in system integration. Ideally, we should be able to do all work error-free on the first pass.

Now we can answer the question, "What is system integration?" It is the rich mixture of three integration components applied in combinations defined by the resultant integration spaces to the work confined to a finite number of integration cells across the program world line that actually comprises system integration on any given program. Throughout the remainder of the book we will consider some of these integration situations as they relate to process and product integration.

4.6 *This may be a little mad*

Prior to leaving this subject we should question one of our assumptions as a way of cementing our understanding of the integration spaces concept. Perhaps we were being premature in dispensing with the INDIVIDUAL EFFORT space. Suppose that Company XYZ has acquired an engineering staff composed of 20 multiple personality individuals and that each of these people has 20 personalities.

Let us say that it will require knowledge of 20 specialized disciplines on Program ABC. Let us also accept that a person with multiple personalities enjoys inter-personality communication of knowledge and that all personalities can be simultaneously functioning. That is, anything known by personality 1 is available to personality 2 and time sharing of the senses and mental processes is not required. We have to accept that speech is a single channel function for these people, however. Finally, the members of the XYZ development staff all graduated from MultiMind University where they have perfected the technique of educating each personality in a different engineering specialty.

Might we not have to re-think the nature of the INDIVIDUAL EFFORT integration space in Company XYZ? One person could completely populate a PDT if the 20 specialties they had mastered happened to match the 20 disciplines needed for Program ABC. Let us assign two engineers to staff the PIT leaving the other 18 to work on 18 simultaneous PDTs. Perhaps this will allow company XYZ to take advantage of the ultimate downsizing opportunity while also accomplishing the work necessary to develop the customer's product.

We have said that the need for integration is driven by the need for humans to specialize. In this case, the specialization process has been internalized and while external integration may not be necessary, there must be a means to integrate the information each personality is working with.

Exactly how does integration take place? It takes place through a human interface called communication appealing to our senses of touch, sight, hearing, and, to a lesser extent, smell and taste depending on the product. We specialists must share a common information base permitting communication of our ideas and an ability to understand the consequences of information communicated to us in association with our own specialty.

If MultiMind University had fouled up the class schedule and educated each of the 20 personalities of Joan Doakes in a different language, Joan would be hard pressed to take advantage of her formidable internal integration possibilities. A common language, logic, and mathematics form the principal integration instruments at the most primitive level.

Whether we are dealing with graduates of MultiMind U. or Georgia Tech, these people must enjoy a common basis of communication and reasoning through which they can exchange ideas that spring from applying their specialized knowledge base to a problem they share in solving.

chapter five

Information systems and communications

5.1 The critical nature of communications

Successful system integration can only occur when the humans performing the work communicate effectively. This simple idea is the single most important notion in this whole book. It is so because the work must be accomplished by many people, none of whom knows everything. The knowledge and conclusions held by each member of a development team must be shared with other team members. This process of sharing information between humans is called communication. Communications connect the many specialists into a network of ideas that all can tap and transform into one powerful force.

We seek to build, in our PDTs, a series of subsystems, not unlike those of the product system we intend to develop. Our product system will include subsystems that must interface (communicate) with other subsystems to exchange information, and the components of each subsystem must interact in the same fashion. We seek to weld the humans working on the PDTs into a temporary system as well. In Chapter 12, we will return to this notion and explain the second most important point in the book, that there should be a strong correlation between the product system composed of things and the program system composed of PDTs.

The principal mode of communications between humans is language composed of words and symbols, about which we have reached agreement, arranged to express ideas. As we know, these ideas can be expressed as marks on paper or their spoken equivalent. So, communication takes place through our power of speech and sign and senses of hearing and eyesight. It is hard to imagine effective communication appealing to other senses except in the isolated but important case of the blind using tactile sensing. The media that we have found most effective in communicating include: language on paper, speech (direct, recorded, transmitted), and, more recently, with computers using keystroked and video displayed language and even the synthesized spoken word.

The single most important communication channel on a program is probably direct, face-to-face human interchange with verbal language. In order to get full benefit from this mode of communications, with the technology available at the time this book was written, the team must be physically collocated such that members are close together facilitating easy verbal communication. It is possible to get some small benefit from excellent telephonic communication between physically dispersed team members, but there is an added dimension to actual direct conversation where eyesight as well as hearing can come into play.

Unfortunately, it is not possible to develop complex systems using only verbal communications, however. We need records and we need instruments through which to communicate complex ideas. It would not be possible to communicate the design of a rocket engine propellant turbopump from the design community to the manufacturing community using only verbal conversation. The richness of the idea defies simple treatment.

Engineering drawings and reports are needed to communicate the designer's conclusions for translation into manufacturing instructions appropriate to the operator of a machine tool whether it be a human being or a computer program. The requirements even for a valve are too many and too complex to trust to conversation between a buyer and a supplier. Without belaboring the point with a blizzard of examples, there is a valid need for the capture of a tremendous amount of information in the development of a system. This information must be organized and stored in a fashion that it can be retrieved easily and quickly. It must also be protected against unauthorized change, loss or damage.

Those familiar with the great fictional detective Sherlock Holmes, will recall that he felt that a detective should not burden his mind with useless information. Nor should system engineers or the information systems that serve them. We must be careful to select the essential information for retention and allow the useless to be lost in order to ensure that our means of retaining information is not overcome and that the process of retrieving needed information is not made unduly complex.

The big question is, of course, "What is important and who decides?" Clearly, the customer has a big stake in what is retained. A DoD customer will define their formal needs in a contract data requirements list (CDRL). For other data that exists, the customer may require the contractor to list it in a data accession list (DAL) and share the list with them. For an agreed upon fee the customer is allowed to obtain any item on the list. The things on the DAL should be every significant data item that is not a CDRL item. The criteria for significance could be a matter of judgment on the part of the program data manager or defined by a written program DAL criteria. A partial list for such a criteria follows (all of these items are listed conditionally on not being CDRL items). The reader will be able to add many other items to the list and should do so as an exercise.

a. Procurement specifications and in-house requirements documents none of which are configuration item specifications so are not CDRL items
b. Internal design review meeting minutes
c. Supplier correspondence and supplier data requirements list (SDRL) items
d. Engineering drawings
e. Analysis reports
f. Test requirements documents, test plans, test procedures, and test reports
g. Program memos
h. Manufacturing planning and stamped planning cards
i. Quality inspection data including deficiency reports and responses
j. Lower tier program plans (assuming the top tier are CDRL items) and procedures
k. Generic company procedures that apply to the program
l. Program plans, controls, and schedules
m. Design decision traceability database
n. Requirements traceability data

Many program managers have agonized over whether particular items should be placed on the DAL due to their sensitive nature. An internal memo might indicate a serious flaw in the design that would be damaging to the company's position with the customer and even end up costing money when there is already a thin profit margin. It is very hard, under these circumstances, to do the right thing. It is very hard to accept that it is in your long term best interest as a company and a person of integrity to provide this kind of information to the customer. However, it is true. A habit of openness with a customer will generally be repaid, if not in this life (program) then in the next. Any deviation from this position requires you to keep two sets of books with the ever present possibility that the second set will come out in a very damaging fashion during the subsequent investigation or litigation over the consequences of the design flaw. There are, of course, sensible as well as self-destructive ways to initially inform a customer about the content of a damaging memo or report.

This kind of memo contains valuable design rationale information that completes the design decision trail for future research on the same product or another. One of the side benefits from being open with the customer in defining DAL content in the broadest possible way is that the combination of CDRL data, DAL data (including SDRL), and contract correspondence defines the materials that should be available in the program library for internal access by everyone within the bounds of security prescribed by the customer for the program. It is possible that another program library category could be identified for proper company private information dealing with competition sensitive information of value in the next phase competition, such as marketing reports. But this category should not be extended to accommodate a place to hide cover-ups of our mistakes.

Okay, so we have decided that we will retain everything of lasting value and we have a criteria for deciding what is important. Now we need to decide in what media we shall retain the data. Those familiar with the data management field understand that the volume of information just from suppliers can become overwhelming in a paper media. The Data Manager on the General Dynamics Space Systems Division Titan-Centaur Program had to move into the electronic age when the paper-filled file cabinets were about to literally push the desks of the data management people out of the available space. The obvious alternative today is to place all data in electronic media and make it available to all via networked computer workstations.

Once you are in electronic media, whether you arrived there driven by physical space limitations or a conscious plan, you will find that many new possibilities present themselves that might not have occurred to you in the good old paper days. In the data management example, instead of physically routing supplier data in serial fashion through your departments for review you will find that it is possible to send it in parallel to all reviewers via the network with no reproduction and distribution delays.

Communication is the exchange of information and, while the most important communication that takes place on a program is direct verbal conversation, the prepared material that will become part of the program library is an essential part of the complete program communication picture. Information can become lost in our paper filing system and the same can happen in our electronic equivalent. So, not only do we need to capture information for future use but we need to be able to access it on demand when that future time arrives.

5.2 A common database

Some companies have made serious and costly attempts to establish a database that contains every piece of information related to their business and product line. One term to describe this entity is a common database. Few, if any, of these efforts have been successful. It is simply very hard to know at the macro level what information should be retained and how all of that information should be interrelated. Most companies have settled on a less ambitious approach involving growth to a common database condition through evolution of databases useful for subsets of the total set of information organized about their process or functional departments. As time progresses, interfaces between these subset databases, initially accomplished by human communication and sneaker networking, become automated leading to an increasingly grand and complete database.

Clearly, the medium to use for information systems involves computers and computer networks. The original computer systems featured very large computers operated by specialists in a central facility in batch mode. People who wished to use these machines had to bring their work to the central point, in some cases in the form of a box of punched cards. The development of microcomputers, powerful engineering workstations, and effective networking hardware and software have radically changed the way information systems are organized. Today it is not uncommon to find Macintosh and IBM-compatible machines distributed within an organization where the working people are. Many companies have interconnected these machines with powerful mainframes and with each other via computer networks enabling a degree of information sharing. Few have taken full advantage of these networks to encourage the degree of information sharing needed to drive integrated product development into the success region.

Is there some useful prescription for employing these resources already available to solve today's information retention and communications problems that also encourages growth toward a final solution? One approach the author has found effective is called an interim common database suggesting that it is on the technology growth path toward a true common database. This approach is also tuned to the dynamic period encountered during the early program phases, where rigid configuration management of data normally is not applied, with a built-in transition to production information retention.

5.3 *Program Interim Common Database (ICDB).*

The ICDB has six fundamental components illustrated in Figure 5-1. A grid of organized storage capability, referred to as a development information grid (DIG), provides a baseline set of information derived from all of the program specialists organized about the product structure. It is a matrix of information with a human interface. This information is accessible by everyone on the program in read-only mode from their networked microcomputer/terminal. It is protected on a network server storage device from being

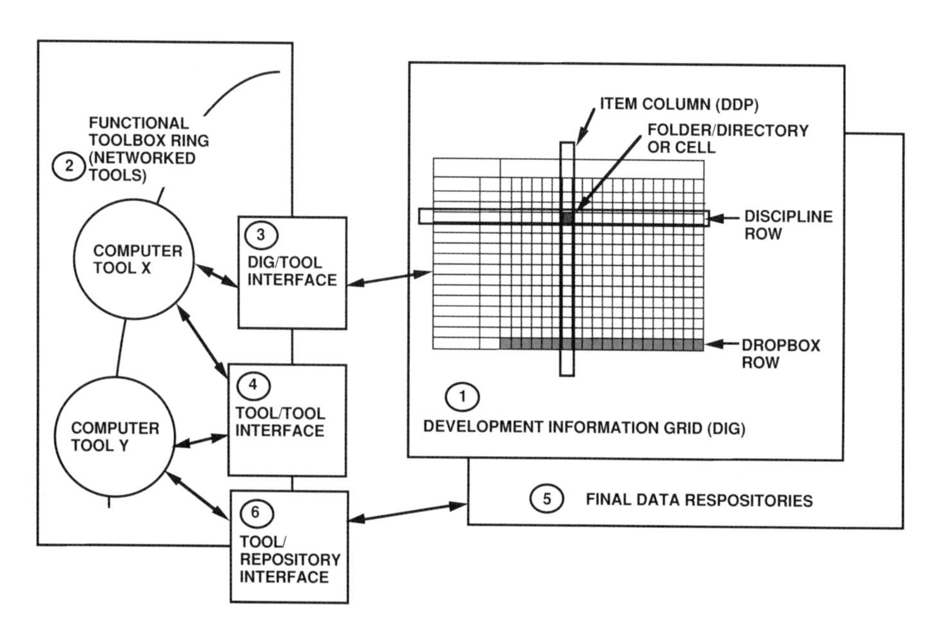

Figure 5-1 Interim Common Database components.

changed except by the assigned expert. The information is entered into and maintained from workstations forming a toolbox ring surrounding the DIG and connected to it via the network. The five components of the ICDB are: (1) the DIG, (2) the toolbox ring workstations, (3) the network resources that interconnect or interface the DIG and toolbox ring, (4) any connections between the tools, (5) the final data repository, and (6) a dataview connection between the toolbox ring and the final repositories.

5.3.1 The DIG

The ICDB centers on a core Development Information Grid (DIG) organized into columns by product architecture (and, therefore, PDTs) and into rows by specialty disciplines as illustrated in Figure 5-2. The vertical columns are called Development Data Packages (DDP) under the responsibility of the PDT Leaders. The horizontal rows correspond to the several specialty disciplines assigned to the teams such as: reliability, mass properties, manufacturing, quality, and so forth. Compare the structure of Figure 5-2 with the structure in Figure 3-2 and you will see a strong correlation between the cells of the DIG and the cells of the organization performing PDT on the program. The DIG may have voids at its intersections as the matrix organization diagram does corresponding to PDTs that do not require particular specialties. The program panes of Figure 3-2 suggests that a company actually needs not one DIG but DIG panes to serve its several programs.

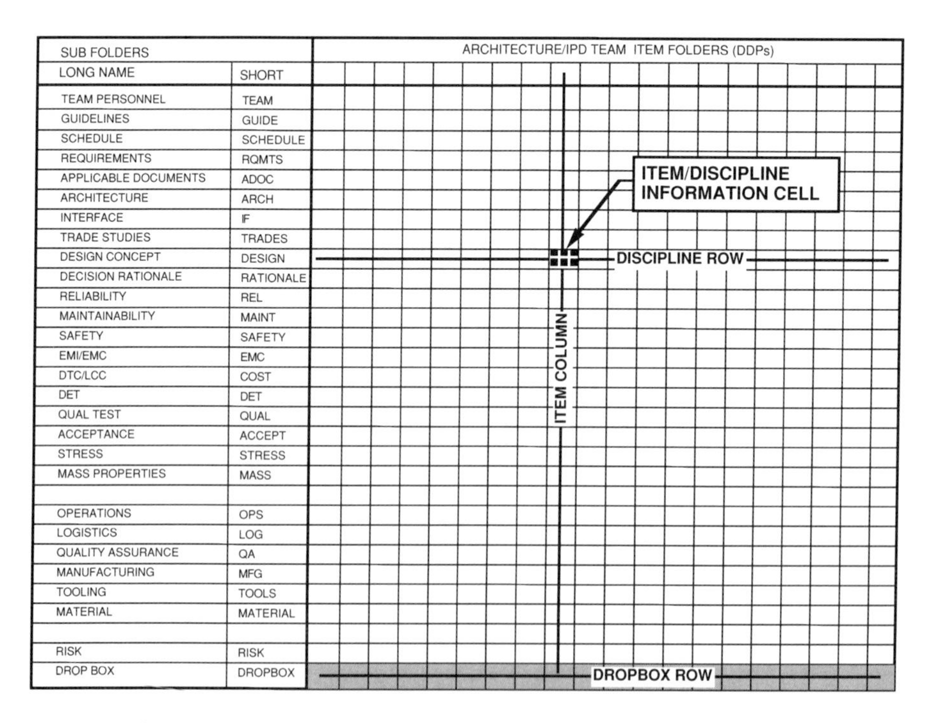

Figure 5-2 Development Information Grid (DIG) organization.

The DDPs are computer folders/directories installed on a computer server not necessarily located within the physical spaces used by the team/project. These share areas are specifically set up for whatever kinds of machines the team has available: MAC only, IBM only, integrated MAC and IBM access, or combinations also involving engineering workstations. Each DDP folder/directory is composed of a column of sub-folders/sub-directories such that everyone on the team has a place to put their team information product. The DDP, therefore, becomes a kind of joint, public engineering notebook for the product item for which the team is responsible.

This engineer's notebook analogy forces recognition of a real problem with the ICDB that must be disposed of as a prerequisite to success. The information in the DIG is accessible to everyone all the time, and that is its strength from a communications perspective. However, many of us humans find it very difficult to share our work product until it is complete and perfect. This attitude is encouraged by management persons who are quicker to criticize than to encourage. Before we inaugurate our ICDB, we must deal with the human issue of access to imperfect and incomplete work. The DIG content should reflect, at any moment, the current information baseline, the best that we know at the time. People will be reluctant to share their evolving work in this open fashion unless they can be assured that it will be a positive experience.

This gets to the very heart of the problem and the solution, to the very fundamental reason for PDTs and the system engineering process. We wish to jointly solve a problem that no one of us can solve. We must share information in order to be successful. We wish to jointly influence each other's work in a true teamwork environment. We wish to work concurrently together to blend our individual knowledge into a force more powerful than the mere sum of our intellects. If we each withhold our work product until it is finished, we will fail to achieve our purpose in organizing this way. If we all make available our evolving work product, some parts of it will influence others to see, yes flaws in that work, but also great new ideas that help us on the road to success. This synergistic effect can build to a critical mass of conversation with an unbelievably strong constructive force. Everything we do to encourage this environment increases the peak of this communication burst and decreases the time (and cost) to get there.

The information contained in the DIG cells may have been placed there by any number of different applications: spreadsheets, wordprocessors, graphics packages, or analytical tools. Naturally, the program, and perhaps the company/division, will be well served to encourage the smallest number of different software products for the same function consistent with permitting the most freedom possible in choosing applications that people can be efficient with. As in so many other things there is a broad continuum here with a large overlapping mutual conflict space. Frankly, some companies have found that it has been advantageous in the long run for them to have allowed employees and departments a degree of freedom in the past in selecting applications with which they could be efficient. It has forced them to become more expert in data connectivity skills than they might otherwise have become. But, ideally it is probably advantageous to work toward a condition of minimized applications with maximum interoperability. All windows and MAC applications can share information via a clipboard, for example. There are an increasing number of applications that can also share data across the IBM/MAC valley.

The immediate past suggests that the future of computer software will continue to be characterized by dynamic change. Some companies have attempted to adopt rigid software application standards for wordprocessing, spread sheet, and graphics applications only to find that other applications arriving on the market were more effective. The U.S. Air Force and NASA managers of the National Launch System program in the late 1980s and early 1990s encouraged an excellent standard for software that we could all benefit from. They encouraged the several contractors, competing in early stages of the program for later hardware contracts, to evolve program and product information system components that consisted to the maximum possible degree of off-the-shelf or over-the-counter software modules that could be replaced easily as newer, higher performance software became available.

Clearly, there is no magic solution in the near term to the question of the right hardware and software upon which to base an ICDB tool ring. You simply have to begin where you are this moment and adopt a continuous process improvement approach to making progress toward a set of goals for your information system needs. It is possible to enter into a condition of paralysis on this matter that is driven by a fear that anything you purchase now will be dated by new improvements very soon. Therefore you should wait for another few weeks which turn into months with no change in the situation. We have all agonized over this effect when deciding on the right time to buy a personal computer.

The author finally understood that the reason he was reluctant to start using a personal computer at work in the early 80s was that he was afraid he would look as stupid as he felt about these machines. He realized that he would have to buy one and experiment with it at home to overcome this inhibition. It was months before that happened, however, because of concern that whatever he bought would be immediately outdated. He finally bought an Apple IIe. Very soon there were a host of better machines on the market. The author finally graduated to an IBM-PS2 with a 286 chip upon which this book was originally written. The 286 chip was way behind the times immediately, of course. But, in this dynamic evolution of computer resources that we will be trapped within for a long long time, you have to put your foot in the water at some point and grow as fast as you can thereafter. It is not possible to intellectualize this process; you must be a participant.

5.3.2 *The toolbox ring*

The computer applications that your enterprise very likely already owns, like word processors, spreadsheets, graphics packages, and specialized analysis tools, reside on the toolbox ring machines. The toolbox ring is formed simply by having them on a network and saying, "Hey machines, you are a toolbox ring." These machines are used to create and maintain program information initially populating the DIG. The engineer uses a particular tool to create a data file and copies the data file into an assigned cell of the DIG via the network interface. This information is immediately available to everyone on the program in read-only mode from all other workstations on the toolbox ring.

The machines need not all be the same type nor must the software be standardized initially. You can simply start with what you have. The machines must be networked together, however. You should work toward an early point where the software does follow some minimal standards to assure maximum access and use of the information. It is possible to configure servers so that both IBM and MAC machines can access the same data. Someone can write a report on a MAC in Microsoft Word and some one on an IBM call it up for review in Microsoft Word, even cut a portion of it and paste it into a different document they are preparing.

There is no science to creating the toolbox ring. You can start, if your machines are already networked, by simply inventorying the machines you have and the software they are licensed to operate. Otherwise, you need to first introduce networking. Then you must decide what software you feel you should be using for each task that commonly has to be accomplished on programs and comparing that with what you have in each application type (such as wordprocessing or spreadsheet). The result is an error signal that can drive your efforts in continuous process improvement.

The functional organization should be held accountable for developing or selecting appropriate computer tools for use by their department members on programs and ensuring that their department members are skilled in the use of these tools in accordance with a standard procedure. Over time, the functional organizations should be improving their tool set in concert with other departments with which their tools have information interfaces. Each program will provide opportunities to experiment with slight changes in the department toolbox and these opportunities should be seized as part of the department continuous process improvement process. As new tools are proven on programs, they should become new department toolbox standards for future programs.

5.3.3 Toolbox ring-to-DIG interface

Early in a particular program, the DIG is used to capture requirements and design (product, manufacturing, logistics, and operations) concepts for joint use throughout the project or during some fixed time span. As the program progresses to a more stable condition of the design and plans, information gradually drains from the DIG via the toolbox ring into the final repository databases where rigid configuration management policies restrict the ease of making changes. Team members reach agreements on who is responsible for creating and maintaining each piece of information needed by the team.

Further, team members agree on which team members must be current on the content of the different team DDP folders. The DDP, therefore, becomes a means to communicate information about the product baseline and encourage consensus. Team members approach their concurrent engineering activities from a position of common knowledge of where they have been, where they are at any moment, and where they are going.

For example, early in a program, the designer for the vehicle hydraulic system uses the folder in the designer row and the hydraulic system column of the DIG to retain his/her concept sketches (using a simple tool like Mac Draw, Designer, or Autocad). This information is available to all of the specialty engineering disciplines concurrently as a basis for their analytical work. The specialty engineers prepare their data and load it into their corresponding folders/directories within the hydraulic system column. The manufacturing people working concurrently with the designer, define a manufacturing process for the hydraulic components that will be manufactured and assembled by the company and this is placed it in the manufacturing row folder of the hydraulic system column. Procurement does the same for things that will be purchased.

PDT Leaders reach agreements with team members on who is responsible for each folder/directory for that team. At any moment in time, the DIG contains within its directories/folders the joint conclusions of the PDT responsible for the vehicle hydraulic system. Everyone on the team and on other teams can gain access to any of this information from their workstation. Two human interface modes are suggested for the DIG: (1) open house mode and (2) control mode.

The open house mode is useful on early program phases when there is a great deal of volatility and a real need to initally load and change the data often to reflect changing concepts. The PDT Leader, in the control mode, will allow people free access to the DIG folders corresponding to their area of responsibility in order to quickly load team data into the folders/directories. Once a baseline is established, reviewed, and approved by the Team Leader, Chief Engineer, or PIT, DIG access should be controlled. You reach this point when you become more concerned about unauthorized changes to the data than you are about stifling creativity. This change is accomplished by retaining read-only rights for everyone and restricting writing rights only to the PDT Leader or a person of his/her choosing for their corresponding column of the DIG, that is, their DDP.

Any team member who wishes to change the content of his/her folder after a baseline has been established (control mode) based on the results of new work, must copy the contents of his/her folder/directory to their local workstation over the network, make needed changes using their toolbox, and then copy the folder or file into the appropriate DDP Dropbox as illustrated in Figure 5-3. The Dropbox is a folder/directory that can only be viewed by the Team Leader or his/her designee but one into which anyone may deposit (write) a file. It is set up so that anyone may write but only the authorized person can see into its contents. This is essentially the reverse of the controls on the discipline cells in the DIG.

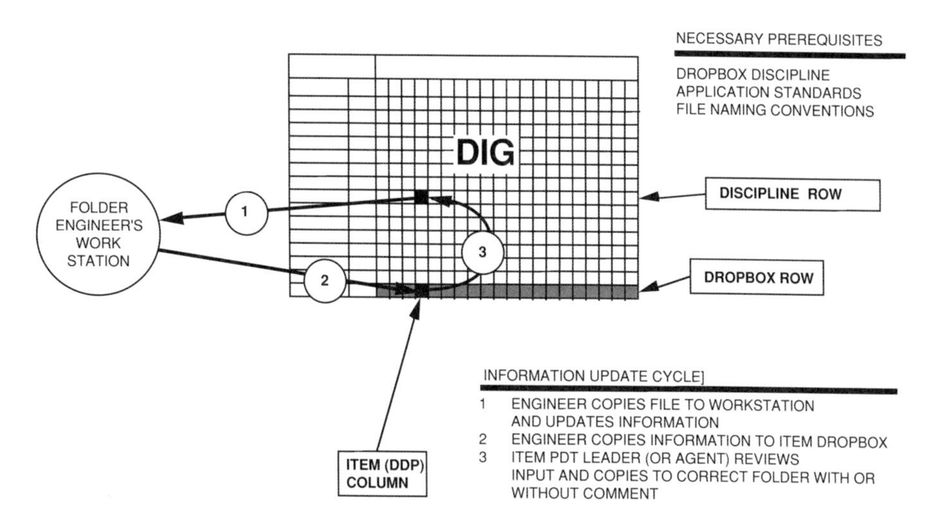

Figure 5-3 Data update in control mode.

The authorized person opens the Dropbox periodically and opens new files deposited in the Dropbox. He/she reviews them for acceptance. Notes may be written upon this material as to its quality, appropriateness, and timeliness and then the file is copied back into the appropriate DDP subfolder from whence it came before modification.

Clearly the DIG depends on a prescribed file naming convention that is strictly enforced. Some operating systems that permit only very short file names, such as DOS up through at least version 5.0, provide a real challenge to make file content and proper DIG cell location obvious from the file names. The Macintosh is much less restricting in this regard and future DOS, Windows, and/or OS2 systems will be as well.

Now, how might the PDT members take advantage of the DIG to support concurrent engineering? The team members can periodically gather around one workstation and discuss DIG content making changes in real time based on the synergism of their conversation. If the team is too large for this and they are physically collocated, they can get together about several workstations in their common work area to view the data in common while carrying on a somewhat louder but no less effective conversation. If the group is too large for this, they can assemble in a meeting room and project the DIG images on the screen making real time changes to content. The projection scheme could also be used in their work area using a clear wall as a projection screen. If some or all of the team members do not share a common facility (some degree of non-collocation) they may gain access to the same imagery for this meeting via the network and converse via a conference call using a speaker phone as the data is changed in accordance with team agreements or team leader decisions. The telephone arrangement will work for hooking up supplier, or even customer, representatives into these concurrent team meetings in any of the alternative meeting styles discussed above provided they are networked with you.

Regular internal team meetings should be held often in a very informal atmosphere with no requirements for preparation of formal presentation materials. All of the needed materials should be available in the DIG. Without a DIG, the only means you will have for the team to interact concurrently is to hold frequent meetings. Without a DIG, the team data must be changed based on meeting results in a two-step process. With a DIG, the data can be refreshed with new ideas during the meeting with everyone in attendance. This is truly a concurrent engineering environment.

With a DIG the team information is available to the team at all times. Without a DIG, the team can only gain access to team information when it calls together the team in a meeting because the team information is largely in the minds of its members. A DIG multiplies the team's opportunities for effective concurrent work because everyone's baseline is available at all times.

These techniques can also obviously be extended to formal customer reviews requiring some degree of preparation, as suggested in Figure 5-4, whether all of the participants are congregated in the contractor's facility or separated by some distance joined through their information and communication systems. The process of creating review data in a computer, only to print paper copies to feed the copy machine producing plastic presentation foils and paper copies is very inefficient. The data can be projected directly from your network with real time changes during the review for relatively simple matters. Other expressed concerns may require assignment of an action item (entered into the network action item database by someone in the

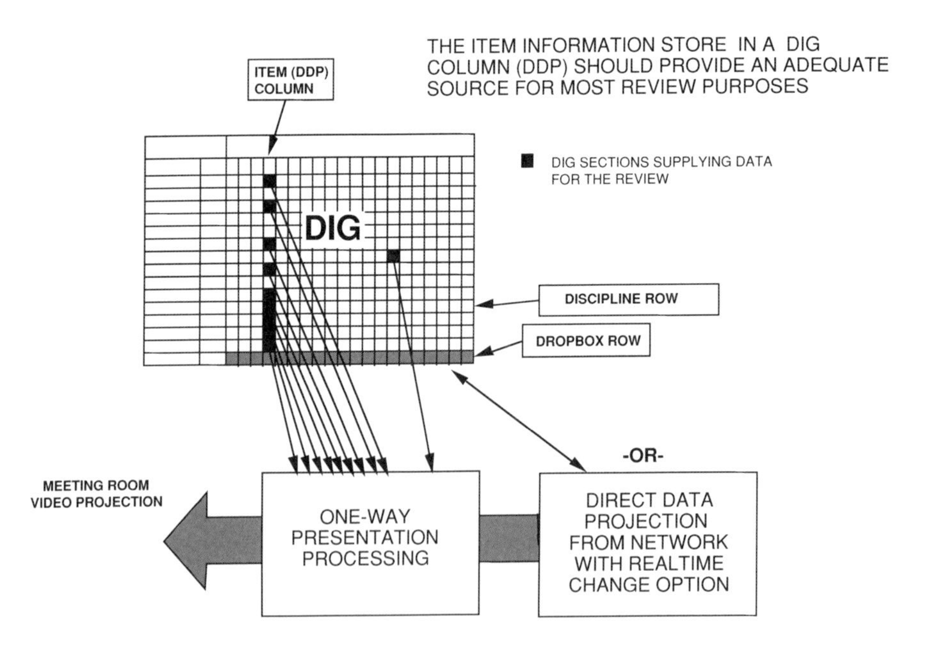

Figure 5-4 The DIG as a meeting presentation source.

meeting room at a meeting support terminal before the action item discussion is complete) and further study.

Alternatively, the planned presentation data could be combined into an integrated data set using a presentation tool like Applause or Powerpoint. This approach permits a sequenced step through of the planned presentation. The network connection in the meeting room also permits the presenter to call up detailed backup data to answer questions about concepts not clearly communicated in the selected presentation materials.

In order to take advantage of these powerful data projection features, meeting rooms must be wired into the network and the presenter's position in meeting rooms must provide access to computer workstation controls or the presenter must be supported by a computer operator. You can select from machines in the range between an expensive high end color projection machine that provides a bright picture in a reasonably well lit room and an inexpensive gray tone overhead projector plate that requires a dimly lit room. Over time, you should evolve to a condition where you have presenter stations with built in computer terminals, possibly with the screen flush with the desk top like those used in TV news anchor stations. In addition, provisions should be made for a computer support station, perhaps in the back of the room. You can see a model of a similar equipment configuration in power software presentations like those put on by Oracle. This would be used for action items, minutes taking, and other record purposes during the meeting. A program or company might have one or more workstations installed on carts that can be wheeled into meeting rooms or used in work areas to satisfy this need.

5.3.4 The tool-to-tool interface

Initially, your ICDB may not permit interaction directly between the toolbox ring tools. As we said at the outset in this Chapter, however, our intent is to use the ICDB concept as a stepping stone toward a true common database. Requirements tools offer an example of how this can be implemented. Figure 5-5 illustrates a requirements tool on the ring that can be updated from a reliability database. The requirements database would have to be structured to assemble the requirements statement from fragments one of which is a quantity field. In the case of reliability, this field would be updated for all items periodically from the reliability model database maintained by the reliability engineer. Reliability requirements data would always reflect the single authoritative current reliability baseline. This arrangement can be extended to weight, quantitative maintainability figures, and other quantitative parameters.

Other examples of possible tool-to-tool interfaces are

a. Traceability relationships between databases for statements of work, product requirements, and WBS Dictionary.
b. An interface between the company personnel database and all the personnel fields in product development databases identifying people

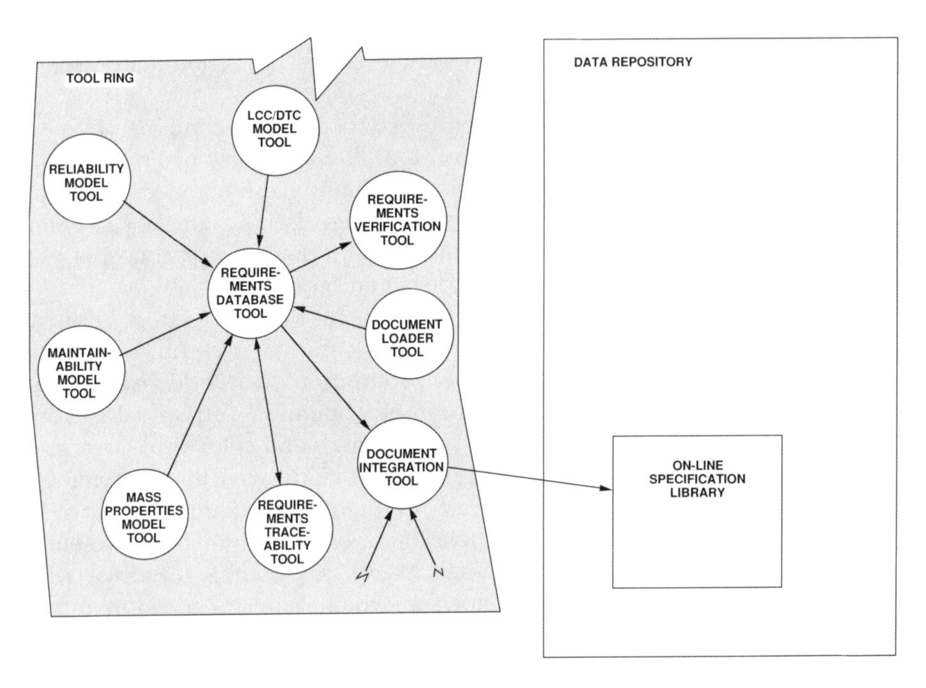

Figure 5-5 Tool-to-tool interfaces.

as principal engineers, actionees on action and issue closure items, and team responsibilities.

c. A product system architecture database that all systems relate their data to. Reliability, mass properties, life cycle cost, and maintainability math models all hook to the same definition of the system such that it is possible to gather the complete story for a given item from databases maintained by different disciplines.

Where a tool-to-tool interface exists, it makes unnecessary the use of a DIG for the data treated this way. The utility of a DIG is in providing an organized repository for data that you have not yet figured out how to relate directly in a common database. As a company progresses in its climb toward a true common database, its need for the DIG should decline during the development period, being replaced gradually by the databases joining the tools on the tool ring.

5.3.5 DIG content evolution to the final data repository

As a program matures, the design sketches in the DIG will be converted into engineering drawings located in a protected Computer-Aided Design (CAD) database and the sketches are no longer needed except as a possible aid in presenting design information at reviews. The manufacturing concepts flow into formal planning data. Procurement data flows into supplier contracts, statements of work, and procurement specifications. Specialty data is transformed into reports conforming to a customer data item description. Requirements database content becomes a pile of specifications. Most of the data in the DIG can be aligned with some terminal form of formal documentation.

Between Preliminary Design Review (PDR) and Critical Design Review (CDR), the content of the DIG should begin flowing out to permanent information residencies. The design concept data in the design sub-folder becomes the source of engineering drawings, the requirements sub-folder content, possibly phrased in primitive terms, evolves into specifications stored in an on-line specification library, and some analysis data matures into CDRL reports stored in an on-line CDRL library.

From CDR onward, the toolbox ring is connected to these final information repositories, rather than the disappearing DIG, with changes regulated by the strict information configuration management rules applied to those systems.

Clearly companies dealing with DoD, DoE, and NASA will have to make this conversion of informal data retained in our ICDB into formal data. But, if we are a commercial firm, the ICDB content may serve our formal needs perfectly and provide us with time-to-market advantages over formal conversion of this data.

5.3.6 Common database approach

As experience is gained in the use of the ICDB, it will be improved over time through continuous process improvement to become a fully networked set of inter-operable real databases from which selected data views may be called by program and functional personnel and, with configuration controls and security provisions, accessible selectively by customer and supplier personnel as well. The tool-to-DIG (item 3 in Figure 5-1) interface will be replaced by the tool-to-tool (item 4 in Figure 5-1) interface and the DIG will wither away being replaced by the final information repositories that the toolbox provides selected dataviews of.

The author has for years been building an integrated toolset he informally calls Rossetta Stone. The real Rosetta Stone included the same message in several languages that allowed us to finally understand one of these languages that had defied deciphering for years. The author's system was intended to capture the information needed by many specialized groups during the system development process and make it available and understandable to all, hence the name. The author uses this environment to experiment with system engineering information concepts to determine what information will be needed in a final common database solution and how that information might have to be interrelated. The requirements components of this toolset are also used in classes taught by the author to demonstrate generic requirements database concepts without endorsing a particular tool

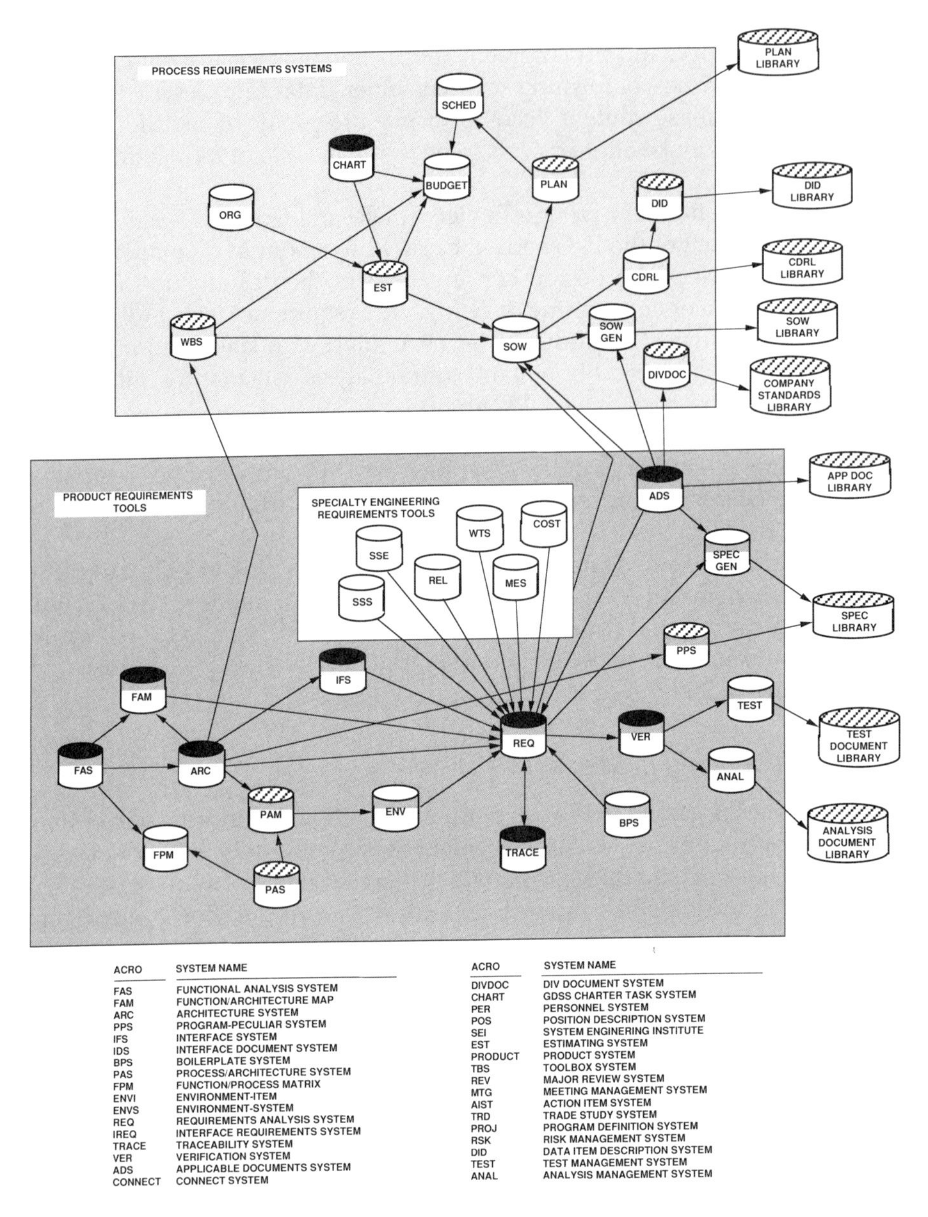

Figure 5-6 Rosetta Stone information relationships.

available on the market. The organization of this experimental system may be of interest as one example of the potential scope of a common database.

Figure 5-6 illustrates the principal interfaces between some of the tools that comprise Rosetta Stone. It shows the top level relationships between the information residing in each system. Each can represents an information system that has one or more companion databases that share certain key, or index, fields with other databases in patterns driven by needed relationships exposed through years of experience in industry.

This system actually now includes 62 component subsystems. Only those related to product and process requirements work are shown in Figure 5-6. Some of these are also supportive of product and process integration work. Other integration and management systems not illustrated include: Risk Management, Project Definition (defines all projects for which the system provides information in one or more subsystems), Configuration Tracking, Major Review, Action Items, Trade Study Management, Personnel, Meeting Management, System Analysis Tracking, Test & Evaluation Tracking, Product Identification, Process Audit, Program Events, Product Representation, and Design Rationale.

The user first selects a program defined in the Project System and then selects an application. This combination defines the computer network server where the corresponding database files are located and the system connects these files to the calling workstation. If the person logging on has authorized access to the server directory within which the files are located, they can change the data, otherwise only read it.

Some systems provide on-line access to document libraries as illustrated on Figure 5-6. The Applicable Document System (ADS), for example, provides legalistic tailoring for several hundred documents but the same database can be used to call the actual text of several documents that have been scanned and corrected. The operator may edit and highlight the text and store it as an in-context tailored document for a specific program. As new products come on line permitting Microsoft Windows access to dBase data (such as Microsoft Access and Foxbase 2.5), the author intends to extend this experiment into the graphical user interfaces (GUI) world.

This whole system uses a character-based human interface presently and the world is rapidly shifting to the GUI model. Many of the systems in Rosetta Stone need to be converted to provide a graphical view port. Functional analysis, state transition diagramming, and schematic block or N-square diagramming are best done by most engineers in a graphical format with hooks to the data. Most serious requirements tools were operating in that environment at the time this book was in manuscript form.

Many available requirements tools provide very effective front end functional analysis capabilities and even simulation features, which are very hard to master but powerful in the hands of a skilled user. Some companies find it very hard to maintain proficiency with this full capability because of the low capture rate on new unprecedented programs. They really need only a small subset of the complex tool as a respository for requirements within which traceability and verification data may be retained and revisions accomplished and from which specifications may be printed.

Some of the tools that provide extensive front end capability do not provide effective specification handling capabilities. Some other tools that provide effective specification data retention do not provide any front end help. The author, after years of exposure to this problem has concluded that a company needs a relatively simple database for retension of specification data, however acquired, that provides good traceability and verification capabilities. If it also provides features for technical performance measurement

(TPM), margin management, specification change notices, and source references, so much the better.

The user access screens for this database should be very simple such that anyone, managers, design engineers, and analysts, can operate it with very little training. This simple system should have an interface with a powerful front end analtical tool on which only a relatively few engineers need to maintain proficiency for performing concept development, functional decomposition, and system simulation work. The results of this work should be ported into the simple database and further requirements work completed in the simple system environment.

The system does include some tentative and unfinished steps in the direction of interoperability. The requirements system (REQ) permits specialty engineering (mass properties, Design To Cost, and reliability, for example) update for quantified requirements such that there is only a single source for the requirement numerical values. Links exist between the Requirements System and the Applicable Documents System for printing Section 2 of a specification with tailoring included. A Development Data System pulls together the complete story expressed in all of the data systems for any given item in the system architecture (ARC).

Most system engineering tools on the market at the time this was written were focused on a single function with minimal interoperability features. There was, however, increasing sensitivity among the tool makers to the market potential for interoperability possibilities. Commonly these first-step features entailed ASCII file exchanges under the guidance of a computer expert. This interface will have to appear seamless to the normal system engineer for it to be fully effective. This will require the development of some very hard earned interface standards respected by the tool makers. The motivation for these standards is that everyone (at least those who produce quality tools) in the tool industry will benefit because there will be an upswing in the use of these kinds of tools. This may be a case of the gas stations on several corners of the same intersection all doing well.

5.4 War room or wall

Each PDT, and the PIT for the whole program, should be provided with a wall or room of walls (ideally with stickboard material applied) upon which they can hang large drawings and materials that help team members visualize team problems and item configuration features that are in work. Even though a team has access to a DIG as described above, there are some things which just cannot easily be grasped from the computer screen no matter how ingenious the computer application. A wall full of the right materials can be very effective in triggering new ideas and insights about team problems. There are stories, it is said in Hollywood, that cannot be properly told on the small screen of television, stories that can only be told on the big screen of a movie theater. The same is true in system development where the big screen is a wall full of information.

A powerful combination is formed by integrating DIG projection with the war room by placing a networked computer with projection capability in the war room. This provides a combination of the grand system view on the walls and any selected detailed view of system information via computer projection. The war room should also be equipped with a speaker phone for remote voice access to meetings held in the room. When the DIG data is fully networked, the speaker phone can expand the size of the war room to include your whole facility where remote engineers on a conference call are looking at the same image on their workstation that is being projected in the war room. This makes for a nice compromise to cover those persons who cannot be physically collocated with the PDT. The PDT and PIT should use a war room for regular standup meetings and maintain the displayed data to support their changing needs.

5.5 *Virtual teams in your future*

If your enterprise intends to attain or retain a position of leadership in your market, you will have to migrate to a condition of information sharing along the lines discussed above in this chapter. The ICDB, or an evolving common database, will also provide exactly what you need to enable a more flexible teaming arrangement than stressed in most of this book in the form of physically collocated teams. Clearly, future teams will consist of people not all physically collocated. As noted at the end of Chapter 3, we will soon be able to effectively interact in a synergistic way even if physically separated through the use of computer resources designed for that purpose.

Virtual teaming will require a common information deposit (our evolving ICDB can serve this need initially) around which our virtual team members may collect for their discussions connected via computers capable of multimedia performance. This will include full motion video of the participants as well as voice and data. These teams will have virtually the same experience as if they were all together in a single war room.

5.6 *Integration excellence = communications*

Integration of product system elements occurs through the human communications patterns within the teams responsible for product development. Everything we do to enhance the communications abilities of the people will appear as benefits in the product features. These benefits will be nowhere more obvious than in the absence of conflict at the product cross-organizational interfaces coinciding with different team responsibilities. Maximize the ability of the people to communicate and minimize their need to do so. This is a sure route to success in system integration and in the development of successful systems.

Part II

Process integration

chapter six

Integrated program planning

6.1 The ultimate requirement and program beginnings

The customer's ultimate requirement is their need, a simple sentence or paragraph that succinctly describes the needed product in terms of its effect on other systems, the environment, or both. This need may be stated by a DoD customer as a new threat that must be defeated, an ordnance transportation and/or delivery demand, or an undersea war-fighting capability, for example. It may also be phrased by a commercial concern about their hoped-for customers. On the surface it would appear that the fundamental difference here is that the commercial company must understand its customer base and their needs well enough to know what will sell whereas the military contractor is told what is needed by their customer in a request for proposal (RFP). Every healthy DoD contractor is in the same boat with the commercial companies actually and must be constantly examining marketing possibilities, developing new ideas that their customer base may find useful, and studying their customer's current situation and potential needs. Also, the DoD contractor increasingly must work to understand the positions and attitudes all of the stakeholders for a given system, possibly even extending to the Nation's population at large in some cases.

So, the phrase "customer need" need not scare away those interested in applying this process to commercial practices. In commercial business it just may be more difficult for the company to find out what the customer's needs are. Success may require good market research, good intuition, lucky guesses, and a very fine pencil for cost and schedule control. Whether the system is of a military or commercial nature, no one need respond to the customer needs, however they become aware of them, with an ineffective process. The planning process included in this chapter can be applied to government or commercial development programs. The first step in this organized process is to understand clearly the customer's need for that is the logical foundation for all other work.

In the structured, top-down development model encouraged in this book, this need is expanded into a system operational requirements document, system requirements document, or system specification through requirements analysis, system modeling and simulation, mission and operations analysis, environmental and environmental impact analyses, and logistics and basing analyses early in program phasing. In addition to the system requirements identified in these early phases, the system architecture is also defined based on the functionality needed to satisfy the need and it is used to fashion a product work breakdown structure (WBS), a hierarchical arrangement of product material and services cost elements. A Department of Defense (DoD) program will use the appropriate MIL-STD-881 appendix as a guide for the WBS but the evolving product architecture based on allocation of needed functionality should be used as the principal input for the WBS. Remember our concerns in this area expressed in Chapter 3. The customer's needed functionality must drive the system architecture, and therefore the WBS, and not the finance community and their inflexible cost-tracking computer implementations.

The program planning work must begin in the proposal period, or early marketing efforts in a commercial situation, and continue during program execution. Where a proposal is required, the management volume or section should reflect much of the content of the program plan. In some cases, the contractor will be required to provide a formal program plan, system engineering management plan (SEMP), or integrated master plan (IMP) with the proposal in draft, preliminary, or final form. In the commercial situation, a new product is a good reason to frame a new development plan or update an old one for new conditions, however simply it may be stated. Granted, there are some commercial product lines richly dependent on serendipity exposed through basic scientific research or good intuition where a degree of non-structure may be conducive to success. But, even in these situations, once the product possibility reaches the conscious mind, its development and distribution can fit into the pattern described here with some tailoring to minimize time to market.

At the time this book was written the world was unwinding from several decades of fierce competition between the East and West involving an intense system development rivalry to develop the unstoppable weapon system and to counter the other side's systems. This competition energized a search for technology that sometimes resulted in new systems requiring a clear sheet of paper approach with several facets pushing the state of the art simultaneously. The systems approach matured from this clean sheet of paper environment, but it can be applied effectively on programs entailing something less such as mild to massive re-engineering or modification of existing systems to accommodate new or changed conditions.

In these cases the system engineering products described in this book may or may not have been created when the system was new or they may have long since been lost or disposed of. So, even if a world class system engineering job had been done on the original system, the results of that work may not be available for use in the changing of that system. In these situations, what can be made clear is the boundaries of the system in terms of its architecture. We can then determine what new functionality is needed and which of the elements of the system architecture must be changed, deleted, or replaced to achieve the new goals set for the changed system.

Once this is known, we can apply essentially the same system engineering process to the development of the affected items as we would apply to a new system. It may be a case of working from the middle out rather than from the top down, but it can be done in an orderly process following the tenets of this book.

At the time this book was written, DoD customers had for years been required to follow MIL-STD-499A in the implementation of a system engineering program. This standard required contractors to develop a SEMP telling how they intended to perform the technical and management activities required by the contract. Very commonly, contractors have written these plans as extensions of the proposal in the same sales language used there rather than as a plan for their own guidance in program execution. This pattern of behavior has been a serious mistake in many companies. Whether a customer requires a management plan or not, your company needs one and it should be written by your company for your company based on a marriage of your customer's needs and your capabilities. This is an integration process, linking customer needs and contractor capabilities, composed of many specialized component parts. At the total plan level it is an example of ALL CROSS integration with an emphasis on the map between functional departments and the development process. Product integration aspects are also included, however.

In this book we will encourage the use of an expansion of a U. S. Air Force initiative referred to as the Integrated Management System. It is a variation of the planning process described in MIL-STD-499B (in review by Department of Defense at the time this book was written) that encourages the development of an Integrated Master Plan rather than a SEMP. The reader interested in commercial markets should not become disinterested because we are going to use a U.S. Air Force model. The fact is that the Air Force has really seized on an excellent planning and management model applicable to most any development situation.

To complete the picture from the contractor's perspective, we will glue onto the Air Force model an internal contractor generic planning capability and continuous process improvement module. We will also merge the planning requirements of MIL-STD-499B and the integrated management system.

In applying the resultant planning system, the contractor or commercial company first must understand themselves, their capabilities and their best practices leading to good customer product value. Next, for each program, they must accomplish a transform between this knowledge of their capability and an expression of the program plan in terms of the customer's need. Then this plan must be well executed. Finally, you must take advantage of lessons learned from each program execution to continuously improve your generic self. This provides an integrated environment within which to attain and maintain a world class capability in your product line. And system integration plays a central role in realizing this goal because we must accomplish our work through the integration of the work products of many specialized people both in the program planning and execution phases.

6.2 Program plan tree

We would all doubtless accept that a specification tree is necessary on a large program to introduce order into the product requirements development effort. What is not as universally accepted is a similar tree for program plans that capture the requirements for the development and production process. All too often programs are implemented allowing planning documentation to be autonomously prepared by the several functional departments (engineering, manufacturing, finance, etc.) contributing work to a program. These plans may be generic company plans or procedure manuals applied to the program or specifically written for the program. Many companies become caught in the trap of trying to be totally responsive to every customer's initially stated process requirements to the extreme that they redesign themselves for each customer in terms of these plans. As we will see in Chapter 10, the future-looking company will apply a continuous process improvement concept to their generic procedures in combination with a rigorous customer procedures tailoring effort on each program to take advantage of the practice-practice-practice notion that world class athletes use on the road to greatness.

Autonomous, functional department planning, disconnected from program requirements, is the target in this chapter. Program process plans should be architected just as the product systems requirements and components should be. Plans should exhibit traceability from the top level plan down through the lower tier plans. A structured, top-down planning process will ensure mutual consistency of all of the program plans with a minimum of surprises during program execution. In the process of preparing program plans, it should not be necessary to come up with a new program design for each proposal and program. It should not be necessary to re-design your company for each customer. We need to find out how to apply the practice-practice-practice technique to our work through careful program planning for specific programs and apply continuous process improvement to our methods with a long term view. Company personnel should be applying the same proven process, incrementally improved in time, to each program and in the process become expert in their specialized disciplines. We return to this concept in Chapter 10.

Very little of the business that a company tries to gain through proposal or marketing efforts involves radically new initiatives. Most of our energy is applied to prospects that are close to our historical product line. Therefore, most of a company's procedures and plans should apply in any new program. This is especially true if your company already does apply an energetic continuous process improvement program.

Given that we are organized in a matrix structure, our functional departments should have procedures covering how they perform their function on programs as members of cross-functional teams. The program must knit the functional methods into a coherent process appropriate for the particular product system under development as appropriate to the development phase.

In this process, programs should not be allowed to substitute alternative processes creatively without acceptance by functional management because the functional departments should be deploying the very best methods they have developed over time based on continuous improvements fed by lessons learned from prior program experiences.

There are, however, two sound reasons for permitting programs to deviate from the current best practices. First, it is through program implementation that improvements can be developed and tested. A particular program may be asked to experiment with a particular technique in defining interfaces, for example. Perhaps the company's history is to use schematic block diagrams and the proposition is that program XYZ will use N-square diagrams instead in combination with a new and promising computer tool that the customer will make available at no cost.

The second reason is that the customer may have a valid need for a job to be done differently than our current best practices cover. Perhaps the

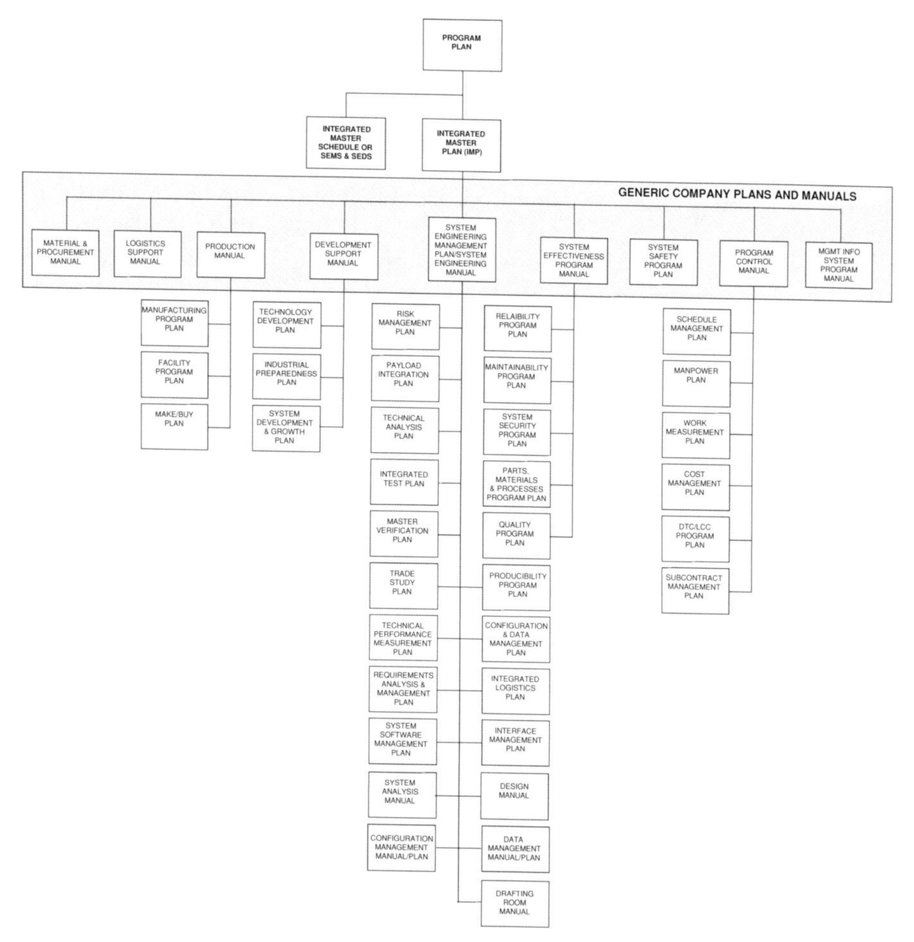

Figure 6-1 Program plan tree example.

company uses a particular computer program and related procedures to capture logistics support analysis data. The customer may have a big investment in capturing the data in a different computer data structure and have a perfectly valid reason why they need to collect different data than your system will address. You will simply have to adjust your practices to the customer's in this case. In the process of doing so, you may find improvements that can be woven into your preferred practice. But, the suggestion is that this should be the exception and the rule should be to follow internal procedures while incrementally improving them for reasons supportable based on prior or current work.

Given that we all accept that programs should be conducted in accordance with prepared plans for each activity, what plans are needed? At the top of this set of program plans rests the program plan. The program plan can be very simple giving the overall schedule in very broad terms, the customer's need, ground rules and policy, top level program organization and responsibilities, and reference to other documentation. Figure 6-1 is based on the integrated managed system but includes a SEMP and subordinate plans to satisfy the IMP requirement for narrative material covering certain planning areas. The Figure 6-1 plan set will satisfy MIL-STD-499B requirements as well as the integrated management approach. The indicated plans provide requirements for the process that will result in the product system. They should be mutually consistent and this can be demonstrated by establishing traceability between the plans in the patterns suggested by the plan tree.

These plans, in combination with program schedules, tell the humans populating the program what to do, when to do it, and who should do it. Where do these plans come from? What are the right plans to write? Who should prepare them? As mentioned above, we can allow our functional departments to autonomously develop program plans and then try to fit them together during program execution. This is a bottom up or grass roots approach to planning. Alternatively, we could approach program implementation planning as we should approach product system development — systematically from the top down. The remainder of the chapter focuses on this approach.

The fundamental difference between the MIL-STD-499B and integrated management system is that IMP defines the work to be accomplished very rigorously linked to the program statement of work and WBS by a common work identification coding system. The functional plans become detailed narrative descriptions of how the work defined in the IMP will be accomplished. Also, the integrated management system calls for an integrated master schedule (IMS) rather than the system engineering master schedule (SEMS) called for in MIL-STD-499B. The advance made by the integrated management system is to absolutely link all work defined in the IMP with the schedules in the IMS using a common work identification system. The remainder of the chapter focuses on the integrated management system approach.

If you are forced by a DoD customer's requirements to plan a program using the MIL-STD-499B approach excluding the integrated management system, the IMP is simply removed from Figure 6-1 and the tree collapses to the program plan and the IMS becomes the SEMS. You may conclude from this that the IMP is an extraneous layer of planning, but you are encouraged to finish the chapter before making that conclusion a permanent part of your belief system.

In the integrated management system approach, you would not normally prepare all of the plans subordinate to the IMP. The content of all of those plans could be incorporated as part of the IMP in narrative form. They are separated here for two reasons. First, a company may have to respond strictly to the MIL-STD-499B requirements, so the book should offer people in that situation this alternative. Second, the principal approach offered in this chapter recognizes a planning approach that makes best use of the limited time available during proposals (by using the generic planning data as narratives referenced in the IMP); focuses proposal work on how these data can be blended into a program-specific IMP; and encourages company use of standard procedures, incrementally improved over time, that provide customers best value. You cannot develop the latter unless you are allowed to practice-practice-practice. If you re-design your company for each new contract, you will never realize a single identity and your work force will not be able to benefit from repetition.

Regardless whether your common customer base can be expected to call MIL-STD-499B in your RFPs, you are encouraged to develop the planning approach offered in this book. You should, of course, sound out your current and potential customers on this planning approach and get their feedback on acceptability. Generally, you will find them to be delighted with the thoroughness of your approach.

Please understand that the intent is that all of the generic plans subordinate to the IMP in Figure 6-1 are intended to be available as a function of ongoing functional department planning that defines generically how the various tasks in department charters are to be accomplished. On a given proposal and subsequent program work, that basis should already be in place. The proposal work should focus on fashioning a program plan, IMP and IMS appropriate to the customer's needs drawing on your standard planning data as narratives referenced in the IMP. If you have ten programs in house, each of the ten programs would have its own program plan and IMP but each IMP would reference the same set of functional manuals that tell the people on the programs how to do their tasks.

You may find writing a generic integrated system engineering manual (SEM)/system engineering management plan (SEMP) difficult if you do not have a model. Another book by the author entitled *System Engineering Planning and Enterprise Identity* provides an attached model for this document including a copy on a computer disk. It is structured using the principles of this Chapter such that it may be used as a program SEMP or the system engineering process narrative referenced in a program IMP.

It is true that some customers will be uncomfortable with your generic planning in that you can change it without their approval after a contract has been awarded. Some customers will want review authority on your internal plans and that can be a nightmare when you have multiple customers with very different interests. In such cases you may have to run a copy of a generic plan and re-identify it for specific use on a program. This program-specific copy could thereafter be changed with customer review without causing chaos in your internal documentation. This will force you to deviate to some extent from the practice-practice-practice notion but is not a bad compromise when forced to respect your customer's requirements for review authority on all program procedures. The program adjusted documents should contain the maximum generic content.

The party line in your proposals and in conversations with your customers should be that you are motivated to continuously develop generic procedures that will provide all of your customers with the best possible value. In cases where the needs to satisfy two or more customers are in conflict, you need to consult with those customers as part of the process improvement activity. You should also offer your customers access to your generic internal planning data on the basis that it is proprietary. Let them act as one of the pathways through which you can detect incompatibility using management by exception. One excellent way to do this is to provide internal read-only access to all planning data via computer network throughout your facility and extend this access to your customer base either at their facilities or only at your own.

In any case, the specific work required for each customer's program would be clearly defined in their program-unique and customer-approved program plan, WBS, SOW, IMP and IMS. These documents would reference the generic planning data that only tells how to do these tasks. The principal obstacle in implementing this approach is, of course, that every company has not done a fine job of documenting their current practices. In the author's opinion, this is not a valid basis for rejection of the offered planning method, nor does it represent an impossible barrier. It means that you must begin developing these generic practices now and keep improving them over time. Your competition may yet give you time to improve your performance for they are likely every bit as fouled-up as you are (if that be the case) in this respect.

In the case where a company is organized in a projectized fashion, or only deals with a single product line, you still have to acquire your personnel from the same source as everyone else, from the human race. We are all knowledge limited and are forced to specialize. So, even if you have no matrix and only a project structure in a company with one or more projects, you will have to accomplish program work using specialists whose work patterns should be standardized, one program to another. This standardization can be captured in the kinds of documents shown in Figure 6-1 under the IMP.

6.3 *Know thyself through generic program planning data*

We will return to generic planning in Chapter 10, but we need to introduce the subject here as the source of much of our program planning data. In the ideal situation, our company will have been involved in a continuous process improvement program for some time and we will have developed a generic set of good practices coordinated with our tools, personnel knowledge, and skills base. This data should include: (1) a functional department charter listing all of the tasks for which each functional department is responsible for maintaining and improving company technology and capability, (2) a generic process flow diagram that hooks these tasks into relative time, (3) a task planning sheet for each charter task (to be explained later), and an integrated company System Engineering Manual and generic System Engineering Management Plan (SEM/SEMP).

The book *System Engineering Planning and Enterprise Identity*, which includes a SEM/SEMP model, also includes in appended data a generic process flow diagram (many parts of which appear in the body of this book) and an accompanying coordinated company charter task list. The generic process flow diagram provides a framework into which all of the charter task descriptions fit as pieces into a puzzle. This process diagram is a simple process flow diagram. It could be expressed as a PERT, CPM, or other network diagram; V diagram; generic schedule diagram; or Gantt chart. The time axis could be in some nominal time measure or the complete period of execution equated to 100%. We can picture this diagram as a rubber sheet being stretched or compressed to satisfy a particular customer's needs with some paths being deleted, others possibly added, as a function of the contract.

Figure 6-2 offers an example of generic planning sheet for one of the tasks illustrated on the flow diagram and listed in the task list. It happens to be a reliability task. Much of this information will have to be changed as it is applied to a specific program, in particular the intensity of task application, but it is helpful to have this information as a beginning point. If you have these sheets available as a generic input to the program planning process, you can save time and introduce your company's standards for process quality into the program planning process commonly accomplished during the proposal development period. If you have no enduring generic planning data now, these forms could be filled in initially as part of your next proposal task estimate and later iteratively improved upon in each subsequent proposal process using the incremental continuous process improvement notion.

Some of the charter tasks expressed on the generic process diagram will entail personnel from only one department working alone to complete them. Many tasks will require cooperative effort on the part of personnel from two or more departments as a function of how you are functionally organized and the business in which you are involved. Therefore, the atomic structure

FUNCT. TASK ID	1051822312
DEPARTMENT	834, RAM Engineering
TASK NAME	Reliability Allocations (Task 202)
PROCEDURE REF	Company Procedure 42-8; MIL-STD-756B (tailored); *Reliability Engineering*, ARINC, Prentice Hall, Chapter 6.
OBJECTIVE	Define a quantitative reliability figure for each system item
DESCRIPTION	The analyst studies the planned design concepts provided by the PDT designer(s) and allocates the system or lower tier failure rate to subelements from top to bottom in each architectural layer and branch based on available historical data or predictions determined by the item composition.
SIGNIFICANT ACCOMPLISHMENT	Reliability figures assigned to all system items.
COMPLETION CRITERIA	The task is complete when every configuration item and procured component has been assigned a reliability figure and the complete set of data has been checked for internal consistency, customer requirements compliance, and been approved by the PIT.
EVENTS	Model structure complete at SDR, allocations loaded by PDR, and predictions loaded by CDR.
RESOURCES	
TOOLS	Five IBM compatible computers each running RAMEASY, Microsoft Word 5.0, and Microsoft Excel for first two quarters. Thereafter, two machines adequate.
PERSONNEL	At peak work load, six experienced reliability analysts skilled in failure rate prediction and reliability mathematics and familiar with equipment used in missile systems.
FACILITIES	Each analyst requires a standard company engineer's workstation. Personnel will be assigned to PDT with 2 assigned to PIT and four to product teams (1, 3, 4, and 5), one per team during the first 2 quarters of the program. Thereafter, two analysts on PIT only.
INPUTS	1. Design concept information from item designer 2. System architecture definition
OUTPUTS	1. Reliability Model complete with a quantitative reliability number for each configuration item and procured item. Data users include: PDT as a source of reliability requirements, Availability analyst as a source of reliability data for computing availability

BUDGET EST PER QUARTER	1	2	3	4	5	6	7	8	9
HEADS	2	5	5	5	2	2	2	2	1
MAN-HOURS	1008	2520	2520	2520	1008	1008	1008	1008	504

Figure 6-2 Sample functional task definition form.

of generic task definition includes a task number and a department (or specialty) number. The generic task described in Figure 6-2 would be referred to as 1051822312-834. This identification number appears quite long due to the number of process layers included, but it is consistent with the generic task definition data in the SEM/SEMP.

The processes that we select for the program become one of the three integration components we discussed in Chapter 4. If two or more people from the same discipline are working in a coordinated way on one task, their work must be co-process integrated.

With generic planning data in hand, we can apply this data in a tinker toy or erector set fashion to particular customer needs expressed in their request for proposal. To do so we need an efficient transform process between generic planning data and specific program planning data. If you look at the planning data prepared in many companies, it includes a mix of functional department and product-oriented data. Our transform process must map the generic planning data into a program and product context understandable to our customers and valuable to us as a basis for managing the program through product-oriented integrated product development teams..

We have assumed throughout this discussion that the company is organized into a matrix structure. You may not be organized in a matrix, but you will still need to focus on the many specialties the technology base associated with your product line demands. The advantage of the matrix is that the functional organization can provide process continuity and a continuous improvement focus while the program organization runs the program. The danger occurs when the functional organization is allowed to enter the program work supervision path.

6.4 *Integrated management system overview*

During the late 1980s and early 1990s the U.S. Air Force, working with several contractors on several major programs including the F-22 Fighter Aircraft Program, evolved an extremely important acquisition management methodology called integrated management system. One of the beautiful aspects of this Air Force integrated planning initiative is that it strips away a buildup of past confusion and shines a brilliant light on the important things in the program planning process. The fundamental notion is that we should first understand the requirements for the product system and then apply a structured approach to design a program that will produce a system compliant with those product requirements. The program design is captured in a system specification, work breakdown structure dictionary, statement of work, a list of deliverable data, a set of program plans, and program scheduling data. The content of the plans are requirements for performance of the process that will produce the product defined by a customer system specification. There should be no work planned or performed on a program that does not contribute in a perfectly clear way to satisfying the customer's product needs expressed in their system specification. This shockingly simple concept is, sadly, not always realized in practice.

This methodology recognizes six fundamental documents that collectively contain the requirements for the product system and the program through which the product will be created. Figure 6-3 illustrates the traceability relationship between these documents from the ultimate requirement, the need. The system specification should be driven by the customer's need and it should recognize a particular top level architecture that is expanded into a work breakdown structure (WBS) dictionary for the purpose of structuring the program for management purposes. A statement of work (SOW) defines the work that must be accomplished for the product system and each WBS element of the system. The required work must then be planned in an integrated master plan (IMP) that fits the work elements into a framework of major program events. Finally, the planned work hooked to major program events is scheduled in time in the integrated master schedule (IMS). The system specification, SOW and IMP all represent the top ends of trees of specifications, statements of work (including supplier SOWs), and plans respectively.

Some large system customers, like DoD, use a device called a contract data requirements list (CDRL) to define their requirements for formal delivery of data about the product and the process for creating it. This document lists each item of data that must be delivered and tells what format it must be in (referencing a data item description, or DID), when it must be delivered, to whom it must be delivered, and how many copies are required. These items are commonly tied to paragraphs in the statement of work and WBS numbers for management purposes. Some of these information products will be part of the delivered product, such as technical orders that customer personnel will use to understand how to operate and maintain the product system. Other data items, such as cost schedule control system reports, meeting minutes, and schedule updates, will be focused on reporting progress in the product system development and manufacturing effort.

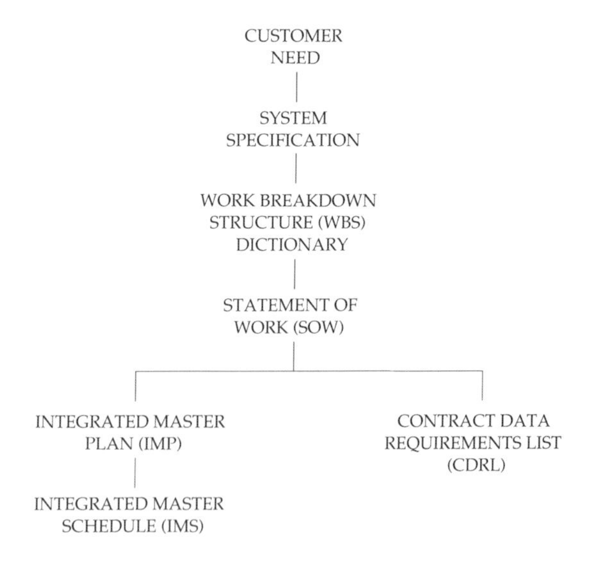

Figure 6-3 Program planning document stream.

6.5 *Generating the six primary documents*

6.5.1 *The system specification*

Depending on the program phase we are discussing, the system specification may be available to us or may not yet exist. Preparing the system specification may be one of the program tasks if the contract involves a very early program phase. More likely, the contractor will receive with a request for proposal some kind of requirements document: such as a system requirements document, operational requirements document, or draft system specification from the customer. This input will have to be completed or transformed into a system specification to the mutual satisfaction of the customer and contractor. Commonly, this transform happens during the proposal process with a system requirements review (SRR) scheduled very early in the program execution. Refer to the companion book *System Requirements Analysis* for a description of a process that will result in a quality system specification.

It often happens that the customer's original need has become lost by the time they are ready to let a contract for the development of their system. Sometimes a need statement is never phrased. But even when it is prepared, it is not uncommon for it to have been so thoroughly digested in the process of developing the system requirements that it simply passes from view. Some people would maintain that once you have the system requirements defined, you do not need one anyway. The author disagrees. The need is the ultimate requirement and, if it truly represents the customer's need, it offers useful guidance throughout the development process.

If you cannot find a need statement in materials provided by the customer, you should ask them for it. If they cannot or will not produce one, write one and get customer acceptance. This may be very difficult, especially where the customer has many faces including one or more users and a procurement agency. But because it is difficult, you should not conclude that it is not needed. Quite the contrary, the harder it is to phrase and gain acceptance of the customer's need, the greater the need to press on to a conclusion. A lot of good work will result that will eliminate many false starts during later program execution. This same policy should be pursued in the definition of system requirements — the harder it is to gain acceptance, the harder you should work to understand the customer's needs and achieve agreement.

In Chapter 11 we will introduce the specification tree overlay of the architecture diagram for which the system specification is the top element. All of the requirements in all of the lower tier product requirements documents should be traceable from and to this document. All of the requirements in the system specification should be traceable to the customer's need through a logical process of functional decomposition, allocation, and requirements analysis.

6.5.2 The work breakdown structure (WBS) dictionary

The WBS has, in the past in the DoD market, all too often been crafted by finance people in industry and government based on one of several MIL-STD-881 appendices reflecting different kinds of systems. Even though MIL-STD-881 encourages that its appendices be used as a guide only, the mindless way it has been applied has often had a chilling effect on the application of a sound system functional decomposition approach to the development of product systems. Despite the evolution of a conflicting functional system architecture on such a program, the inflexible computer cost management tools used by the customer and contractor finance communities has often inhibited alignment of the WBS to the needed functionally derived architecture. Under the Air Force initiative, the government has not shown complete flexibility on WBS adjustment, but the climate is more favorable than in years past. We will return to this problem in Chapter 11.

The WBS Dictionary provides a hierarchical organization of product material and services needed to satisfy the customer's need. The WBS must span the complete system allowing everything in the system to be placed in some WBS category. The content of the WBS must reflect what is needed to satisfy product system functionality and should not be chosen rigidly and arbitrarily based on some financial model. Figure 6-4 lists the partial content of Appendix C of MIL-STD-881B for missile systems. Only the Air Vehicle, a 2nd level WBS element, is expanded to the 3rd level.

MIL-STD-881 does not prescribe a numerical coding system for the structure, but one is always assigned. For example, in Figure 6-4 the Air Vehicle might be assigned WBS 1000 and the other 2nd level items 2000, 3000, and so on. WBS 0000 would be assigned to the whole system in this case. If the system included an Air Vehicle and other things, like a launch site and a final assembly factory, WBS 0000 would be assigned to the complete system and each of these elements, sometimes called segments, would be assigned thousand level WBS codes. Some programs are so complex that prefixes are assigned to permit cost accumulations in different useful patterns. The WBS 012-1000 might be assigned for non-recurring development of the Air Vehicle while 014-1000 is assigned to recurring manufacturing of the Air Vehicle. Costs can be accumulated in WBS 1000 across all prefixes for the complete Air Vehicle cost and within prefix 012 for all development cost.

Lower tier WBS identification is accomplished by using different numbers in the hundreds and/or tens place of the four digit WBS code. For example, WBS 1100 might be assigned to the first stage propulsion system, 1200 to the second stage propulsion system, and 1210 to the rocket engine of the second stage propulsion system. We have chosen a four digit code so far but there is nothing to prevent the selection of a five or six digit code. In a very complex system, a decimal expansion could be added, such as 1210.05 for an engine turbopump. In this fashion, the WBS can identify everything in the product system in a very organized fashion to any level of indenture desired. It is not necessary to apply a unique WBS number to each product

Missile System
- Air Vehicle
  - Propulsion (Stages 1...n, as required)
  - Payload
  - Airframe
  - Reentry System
  - Guidance & Control Equipment
  - Ordnance Initiation Set
  - Airborne Test Equipment
  - Airborne Training Equipment
  - Auxiliary Equipment
  - Integration, Assembly, Test, and Checkout
- Command and Launch Equipment
- System Engineering/Management
- Systems Test and Evaluation
- Training
- Data
- Peculiar Support Equipment
- Operational/Site Activation
- Industrial Facilities
- Initial Spares and Repair Parts

Figure 6-4 Example of a MIL-STD-881 work breakdown structure.

item throughout the system hierarchy, however. The WBS is a management tool and should give managers insight into the system structure. In Chapter 11 we will observe that the WBS is actually (or should be) an overlay on the functionally derived system architecture which should be expanded to the component item level (valves, black boxes, mechanical assemblies, etc.). Not all of this detail is needed for management of the system development.

In Chapter 11 we will uncover some common flaws in the development of the WBS that can result in it becoming disconnected from item functional association and organizational responsibilities for development of the product system elements. For now let us assume that the WBS is properly developed to reflect these views. Let us assume for now in the remainder of this chapter that there is a clear relationship between the WBS and the things the system must do and the assignment of responsibility for development of system items that will accomplish these things.

6.5.3 *The statement of work*

The next step in the programmatic requirements development process is to determine what work must be performed to develop, design, manufacture, test, and deploy every product element depicted in the WBS. Every bit of work we perform on a development contract should be included in the statement of work (SOW) at some level of detail and should be traceable to the product requirements in the system specification. The SOW, therefore, will tell what work must be accomplished for each product WBS element at some level of indenture. Only then can the program work be said to flow from the system requirements.

In the past, the SOW has often been prepared by someone in the customer's program office copying customer-created boilerplate SOW material from a similar past program into the new program SOW. The integrated system management approach calls for the contractor to write the SOW based on a WBS derived from the same work that produced the definition of the system contained in system specification. The contractor must decide what work must be done within the context of his plant, personnel base, and product history in order to create a product system that satisfies the provided requirements and organize the work around the breakdown in the WBS.

Two competing contractors may very well offer the customer two very different statements of work with their proposals because they have different plants, product histories and experiences, and personnel mixes. Neither may necessarily be better than the other, only different for these reasons.

If we have properly identified program work, it should be possible to establish traceability between the paragraphs of the SOW and the paragraphs of the system specification. Where data item description (DID) CMAN 80008A is called by a DoD customer to define the system specification format, the product-oriented SOW paragraphs can be traced to paragraphs under system specification paragraph 3.7, which captures requirements for major items in the system architecture or WBS. Paragraphs under 3.7 should have been initially conceived from an orderly functional decomposition of the customer need and the WBS developed from that same analytical process..

We can picture all of the supplier statements of work strung out from the system SOW in a tree structure branching from the SOW block of Figure 6-3. Certainly traceability should exist between the process requirements we accept in the applicable documents referenced in the statement of work with our customer and the process requirements we lay upon our suppliers in procurement SOWs. Otherwise, some elements of the complete product delivered by us (containing supplier elements) may not be compliant with our customer's requirements. The Air Force planning initiative brings into clear focus the need for traceability not only through the specification tree but through the SOW tree as well and between a SOW and its companion specification. This means that the complete product and process definition for a system can be unfolded from the customer's need in a structured top-down development effort.

Commonly, the system level SOW is written to cover all of the work in several WBS indentures. We prepare supplier statements of work for major suppliers but the system SOW commonly provides the only work definition coverage for the prime contractor. The product development team (PDT), or concurrent engineering, paradigm suggests an interesting alternative to this arrangement. We could prepare the system level SOW to cover only the system level work under the responsibility of the PIT and write internal SOWs for each major system element identified in the WBS and to which a PDT will be assigned. These internal SOWs would each define the work that must be accomplished by one of the teams. The product element would be covered by a specification defining the product requirements and the team

SOW would define the process or work requirements that must be accomplished to satisfy the corresponding product requirements.

This arrangement results in product requirements and work definition documents aligned perfectly with the product system elements and the development responsibility definition and should result in great precision in management of the program. In order to be successful in this approach, you have to have thoroughly studied the customer's need and requirements and decomposed their need into a stable architecture that can be assigned to product teams. Instability in this whole structure can result in a tremendous amount of parasitic work. You can begin this planning work too soon as well as too late.

The SOW, as the name implies, identifies work that must be accomplished at a high level to provide the items identified in the WBS. Now, how do we identify or describe the work that must be done in detail? A three-step process is suggested.

The first step in transforming generic planning data into the specific program plan for a given program is to map the generic tasks to the WBS. Figure 6-5 illustrates this process. It can be accomplished in a top-down or bottom-up fashion and be accomplished at any level of management desired either by the program team or functional management. Two alternatives for these three planning process components are listed below under headings in the form Development Direction/Planning Level/Planner and the reader can imagine other possible combinations:

a. Bottom-Up/Department Chief Level/Chief — Each functional department Chief, or their representative, lists under each WBS, their department tasks that must be performed.
b. Top-Down/Upper Management Level/Program Planning Team — A program planning team, composed of people from functional departments, accomplishes the map between the functional department charters and the WBS. The planners shop the functional charters for tasks that satisfy the customer's needs expressed by the WBS Dictionary.

Step 2 is to organize the functional task inputs into a program context within the WBS-driven SOW hierarchy. In the author's preferred scenario, no matter how we orchestrated the first step above, the program should form an integrated planning team composed of people from the functional departments assigned to work on the proposal or program initiation work. Ideally, these same people will later be assigned to work on the actual program from those departments.

The planning work for each WBS should be led by the person who will be responsible during program execution for that WBS, if possible. In the case of product-oriented WBS elements, at some level of indenture this should be a PDT leader. The planning team members for a given WBS must study the functional charter task sheets (see the sample in Figure 6-2) assigned to their WBS and organize, integrate, or synthesize them into a specific set of major

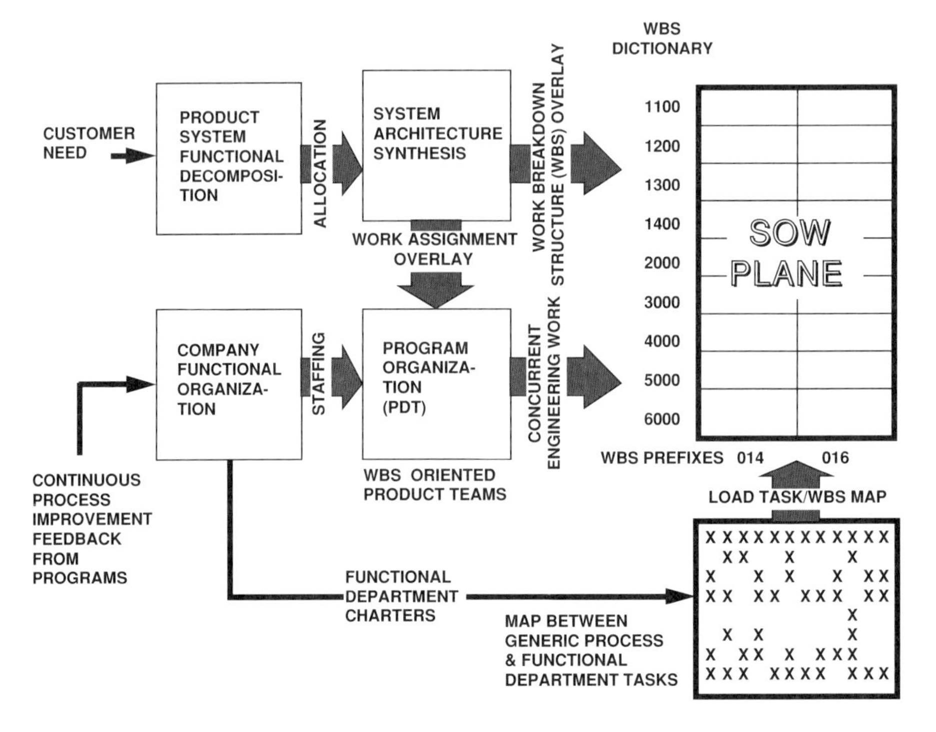

Figure 6-5 Initial SOW task loading.

program tasks that must be accomplished to satisfy the work element goal or purpose. The aggregate results of this work forms the SOW plane of the IMP Work Space shown in Figure 6-6. We might conclude, as a result of this planning exercise, that all of the work that must be done for WBS 014-1200 (prefixes not necessarily used in a specific situation) can be achieved within the context of major program SOW tasks 014-1200-01 through 014-1200-07, 016-1200-01, and 016-1200-02, for example.

Some of the detailed functional task sheet inputs may be gold plated, even unnecessary empire-building attempts. Some may be understated or even missing. They must be assembled into the fabric of the program by people experienced in the company's product line, organizational structure, and functional department charters and the needs of the customer. This is an application of ALL CROSS integration seeking out a condition of minimized completeness.

We have to assure ourselves that we have covered all of the necessary work and have not introduced any unnecessary work. For each planning sheet offered, we have to ask, "What would happen if we did not do this work or did it at a lower level of intensity?" For each WBS element we have to assemble the complete stack of planning sheets (physically or in our computer) and ask, "Have we covered everything that needs to be done to satisfy this work element?" There is, of course, no substitute for successful program planning experience in this integration process no matter what kind of whiz-bang computer system you are fortunate enough to have.

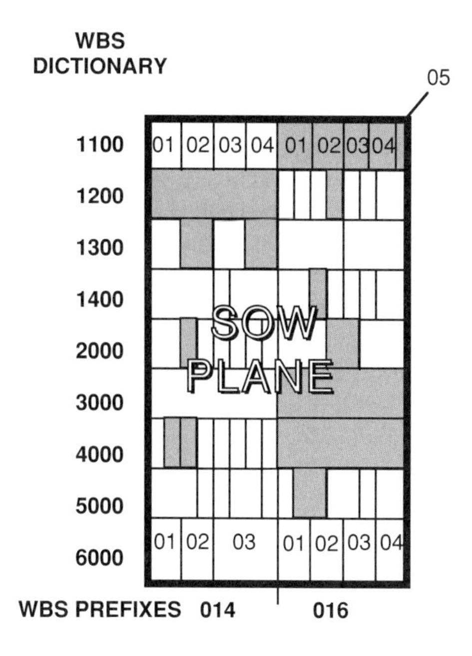

Figure 6-6 The SOW task plane.

Prior to kicking off this SOW task planning exercise, the WBS should be well developed and made available to everyone involved in the work planning exercise. It is a great source of confusion when the WBS is being changed twice a day in the middle of the work planning activity. Of course, the WBS may have to change based on new knowledge developed in this process, but the more stability that can be introduced into the WBS prior to beginning the more detailed SOW planning work the better.

Very little of the planning information shown in Figure 6-2 need be included in the SOW, but it is needed for the overall planning process including budget determination (cost estimating), detailed program planning, and program facilitization, staffing, and equipping. Planning sheets like that shown in Figure 6-2 received from the detailed task planners form the atomic structure of the detailed planning activity that will result in the IMP and IMS. These inputs must first be organized within the WBS, as discussed above, and woven into an understandable presentation in the SOW. Figure 6-7 includes a fragment of the SOW framed in the integrated system management style organized not by paragraph numbers but by an expanding work ID number formed from the WBS prefix (if used), WBS number, and SOW Task number.

These SOW tasks are fairly generic in nature and are appropriate for a wide range of programs. Note that in the two alternative mapping processes we discussed above you could have used a top down list like this created by the program prior to the functional department loading of the SOW tasks. The alternative is that you allow the functional departments to load their tasks into the WBS numbers and integrate the result to gain insight into the

SOW tasks by combining and unifying. The author prefers the former method but no matter how you gained insight into them, they are the program tasks to which we have mapped all of the functional charter tasks.

Twenty different departments may have mapped a charter task to SOW task 014-1200-01. A system engineering department could have indicated they will coordinate the requirements development process for the item/team. Fifteen different specialty engineering departments (reliability, maintainability, mass properties, design-to-cost, etc.) may have indicated they must develop one or more requirements for the item. Functional charter task 105182312-834, displayed in Figure 6-2, could be one of these. A manufacturing engineering organization and logistics department will both cooperate in development of requirements that are consistent with manufacturing and system operation needs. A test & evaluation department will have to define verification requirements. And finally, an ops analysis department may have

014	Non-recurring work
014-1200	Non-recurring upper stage development work.
014-1200-01	Define upper stage requirements. Conduct analyses to define appropriate requirements for the upper stage based on system requirements and the evolving system concept.
014-1200-02	Review and approve upper stage requirements. Subject the requirements to formal review and approval prior to design authorization.
014-1200-03	Design upper stage. Develop a preliminary and detailed design fully compliant with item requirements.
014-1200-04	Provide design test and analysis support. Perform analyses and tests for the purpose of validating the adequacy of the design concept.
014-1200-05	Review and approve design. Subject the design in preliminary and detailed stages of maturity to a formal review by qualified specialists and approve same with possible re-direction.
014-1200-06	Verify design by analysis. Analyze upper stage design for compliance with requirements.
014-1200-07	Verify design by test. Conduct testing to verify that the design complies with item requirements.
014-1200-08	Identify material sources. Define the source of each item of material needed to manufacture the item (make/buy).
014-1200-09	Develop manufacturing process. Define facilitization, tooling and test equipment, and personnel needs.
014-1200-10	Develop inspection process. Define inspection steps necessary in coordination with manufacturing process definition.
016-1200	Recurring upper stage work.
016-1200-01	Acquire and distribute material. Execute material acquisition plans, stock received material, and make available to the production process.
016-1200-02	Manufacture, assemble, and test upper stage. Perform all necessary steps to assemble materials defined in engineering drawings in accordance with manufacturing planning and inspection data. Test upper stage in accordance with approved written instructions for the purpose of assuring item readiness for acceptance by the customer.
016-1200-02	Deliver upper stage. Package and ship item to launch site.

Figure 6-7 SOW sample fragment.

to perform a special simulation to define the most cost effective guidance accuracy allocation for the item. Keep in mind through all of this planning work that, while we discuss a mapping of work to functional departments, the work will actually be accomplished on programs by integrated teams of specialized engineers from these departments collocated in program spaces and managed by the programs.

In the process of combining all of these functional department charter tasks into SOW task 014-1200-01, we have also defined the specialists who must be brought together in a team environment to concurrently develop the requirements for the item. The WBS leader/PDT leader who performed the task integration process will have derived knowledge of his/her team composition as a function of the planing experience. Table 6-1 illustrates a fragment of a SOW task responsibility matrix that we could now begin to build. The matrix will be useful in controlling program implementation. Each SOW task is identified by number and name (using a subset of those identified in Figure 6-7) followed by the task status (STAT) column, PIT/PDT responsibilities definition, and needed specialized functional department participation. Team 1, in this example, is the PIT and the other teams are PDTs. The functioanl department information tells where to get the people to perform the tasks. Given that we have progressed in the program to the point that requirements are complete and approved, the "C" in the STAT column would convey that condition. The "A" in the design row would indicate the team may perform design work and supporting analysis and test work.

6.5.4 Integrated master plan and schedule

Now that we know how the product system and development process will be organized (WBS) and what work must be done (SOW content expanded by the detailed planning sheets mapped to the SOW tasks), it is necessary to determine who will be responsible (not in the functional organization sense but within the program PDT structure), how it will be done, and when it will be done. We could establish a tree of plans to define these things like that illustrated in Figure 6-1a topped by a program plan flowing down to a system engineering management plan (SEMP) and sub-plans such as: manufacturing plan, procurement plan, quality assurance plan, etc.

The Air Force integrated management system initiative calls for an integrated master plan (IMP) and that is the pattern we will follow here. This does not mean that a program team should overlook all the other plans shown on Figure 6-1. Many of those are still appropriate as IMP subplans or integral IMP narrative content. The IMP can replace the program plan and even the SEMP but may not provide the detail needed by manufacturing, quality, and many other functions. The integrated management system initiative recognizes that all IMP content does not fit neatly into the highly organized structure explained here. It also calls for narrative sections where appropriate. The content of the functional lower tier plans, if prepared can provide the narrative data encouraged in the IMP.

Table 6-1 Task responsibility matrix fragment.

			TEAM			FUNCTIONAL DEPT						
SOW TASK	TASK NAME	STAT	1	2	3	A	B	C	D	E	F	G
014-1200-01	Define upper stage requirements	C			X			X	X		X	
014-1200-02	Review and approve upper stage requirements	C	X				X					
014-1200-03	Design upper stage	A			X		X		X	X	X	X
014-1200-04	Provide design test and analysis support	A			X							
014-1200-05	Review and approve design		X				X					
014-1200-06	Verify design by analysis				X			X		X		
014-1200-07	Verify design by test				X		X					
014-1200-08	Identify material sources				X				X			
014-1200-09	Develop manufacturing process				X						X	
014-1200-10	Develop inspection process				X							X

The author proposes tying the MIL-STD-499B prescription together with the integrated management system initiative by recognizing that the SEMP is essentially the system engineering narrative of the IMP. Other functional plans can have the same relationship to other IMP narratives. The author's idea of the ideal situation is that the contractors should have a set of functional department plans or manuals that describe how their chapter tasks are to be accomplished. One of these plans would be a generic SEMP/system engineering manual that would tell how the company accomplishes the system engineering process on all contracts. If the company deals with the military as a principal part of its business, then its generic SEMP should map to MIL-STD-499B content with possible tailoring clearly defined in any contract calling for a SEMP.

For a given contract/program, the company would write a WBS Dictionary/SOW/ IMP/IMS document set and in the IMP, where the customer requires narrative materials, refer to the lower tier functional plans for detailed coverage. This arrangement allows company personnel to follow the same practices on all programs developing and maintaining skills in a single process. Continuous process improvement based on program lessons learned can then be applied to a sound base to move toward the very best possible capability constrained by available resources.

The IMP expands on the SOW telling how and by whom the work will be accomplished and provides a means to determine successful work completion. For each SOW task, at some level of WBS indenture, the IMP must contain planning data that identifies a series of major events through which the development of that product WBS will be accomplished. These events are selected with effective program management in mind. We do not want so many events that we are spending all of our time reporting. Nor do we want so few that we know not what is going on. This is like selecting test points in a system design that allow rapid identification, isolation, and correction of problems. Too many test points costs money unnecessarily and are tedious to use in isolating faults while too few test points leads to ambiguity in fault isolation. These same extremes apply to program health monitoring systems for management.

6.5.4.1 Program events definition

The principal event discriminator will be defined by a customer or by your marketing department's determination of a potential date that commercial competitors might be ready to market a competing product. In DoD this is called an initial operating capability (IOC) marked by a specific date in the future. This date, or its commercial equivalent, defines a point in time when the program must have completed testing and have produced enough product articles to equip the customer for what they have defined as IOC or to fill anticipated orders. This might correspond to two aircraft squadrons, one armored battalion equipped, or production of the first day's assembly line run of new cars.

Your scheduling experts will then have to fit all program activity in between the program beginning time and IOC with some manufacturing and logistics events extending beyond IOC. A generic process diagram, drawn on a rubber sheet in our imagination, can be useful as noted in Figure 6-8. The major events on the generic diagram can be pushed and pulled into alignment with the realities uncovered by the scheduling experts.

Depending on the customer, they will insist on a particular stream of major reviews, possibly those included in Table 6-2. Other major events, or milestones, unique to your product line can be added based on the kind of product, maturity of the product development process, and degree of intensity with which customer or company management wishes to manage the program. The list may include only a single program phase or recognize the full sweep of the development, deployment, and operational steps in the development portion of the system life cycle.

The list of events in time must now be applied to the IMP space as shown on Figure 6-9. The rubber sheet generic process diagram now can be stretched and compressed to conform with the program-specific events and be used as an aid in transforming the rough SOW task planning results into the fine structure provided by IMP task planning.

All of the SOW tasks now need to be laid into this time axis. The team responsible for each SOW task should expand the planning detail for their activities by linking up each SOW task with the events. This breaks up each SOW task into a number of segments like the sample illustrated in Figure 6-9. Some of the periods of time between events will correspond to voids for some SOW tasks as indicated by shading in Figure 6-9 meaning no work is planned for those periods and tasks.

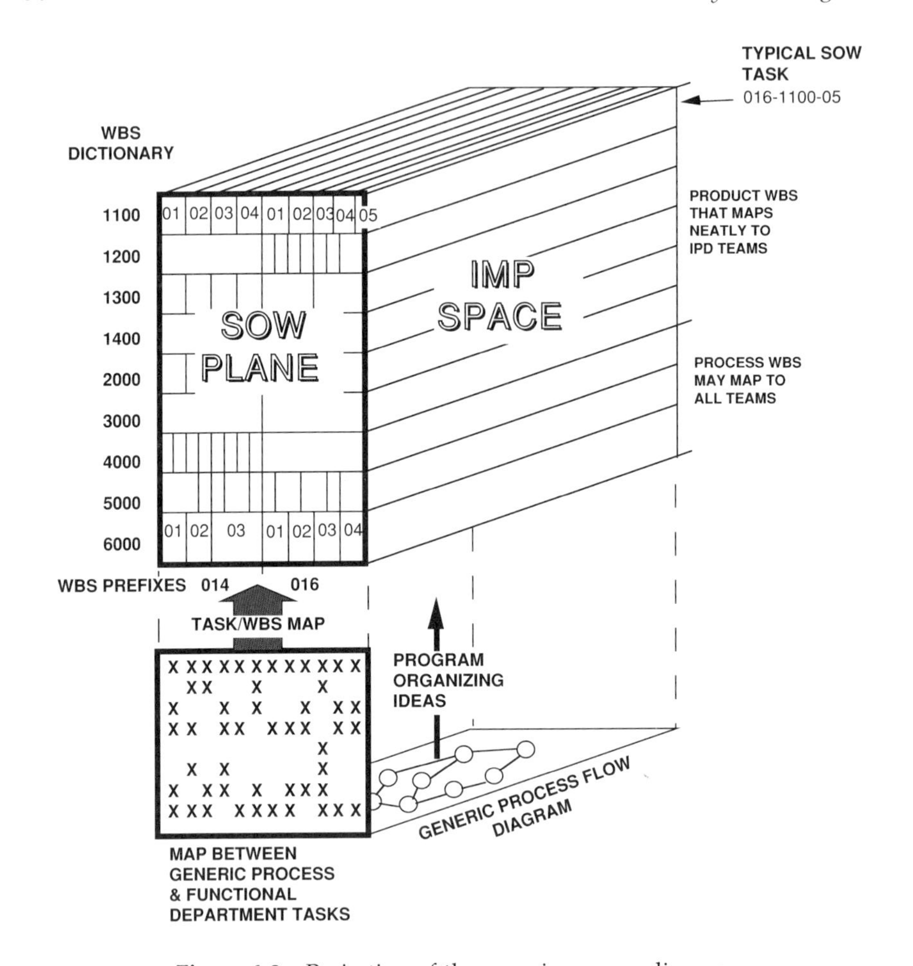

Figure 6-8 Projection of the generic process diagram.

6.5.4.2 Final work definition steps

The work components formed by mapping generic functional tasks into the WBS Dictionary, partitioning and combining the work mapped to specific WBS numbers into major SOW tasks, and the time delimitation of those SOW tasks into events provide us with manageable increments of work that are associated with specific WBS numbers, most of which will map to PDTs in an unambiguous fashion if we have organized the teams properly to reflect the product structure. To enhance the management potential for these tasks, the integrated system management initiative calls upon the planner to identify one or more significant accomplishments for each of these IMP tasks and for each accomplishment to identify a measurable accomplishment criteria through which we can objectively determine when the accomplishment can be claimed.

Figure 6-10 illustrates these final IMP task definition parameters. An IMP task is a work increment between two events within a given SOW task. It is

Table 6-2 Major program event list.

ID	Acro	Event Name	Event Description
01	SRR	System Requirements Review	Joint understanding of requirements reached.
02	SDR	System Design Review	System requirements validated with risks identified and mitigated to an acceptable level.
03	IRR	Item Requirements Review	Approval of item requirements as a prerequisite to item design work.
04	PDR	Item Preliminary Design Review	Acceptance of readiness to undertake detailed design.
05	PDR	System Preliminary Design Review	All item PDRs complete and acceptance of detailed design entry.
06	CDR	Item Critical Design Review	Design acceptance and authority to initiate manufacturing of test articles or low rate production.
07	CDR	System Critical Design Review	All item CDRs complete.
08	FCA	Functional Configuration Audit	Proof that design will satisfy requirements.
09	PCA	Physical Configuration Audit	Proof that product will comply with design, quality, and manufacturing planning.
10	IOC	Initial Operating Capability	Sufficient resources manufactured and delivered to provide a complete operating capability at some level of service.

composed of one or more subtasks each of which is characterized by a significant accomplishment statement. Each significant accomplishment must have identified for it one or more unambiguous accomplishment criteria.

Table 6-3 illustrates one simple string in this planning process using a dash delimited numbering system. Paragraph 3.7.1.2 of the system specification calls for a space launch vehicle upper stage to accomplish some previously defined functionality. WBS 014-1200 was selected to identify upper stage development work. The SOW has expanded on WBS 014-1200 to include several work tasks needed to satisfy the WBS 014-1200 requirement. Based on an analysis of the evolving IMS (or SEMS), we selected completion of design as a critical event through which to manage the program such that we could review design progress before we committed to manufacture of the flight test vehicles. Traditionally in the DoD environment, this event is called the Critical Design Review (CDR). We will accept that the launch vehicle design is complete if 95% of all planned engineering drawings have been formally released at that time. Scheduling, with the concurrence of program

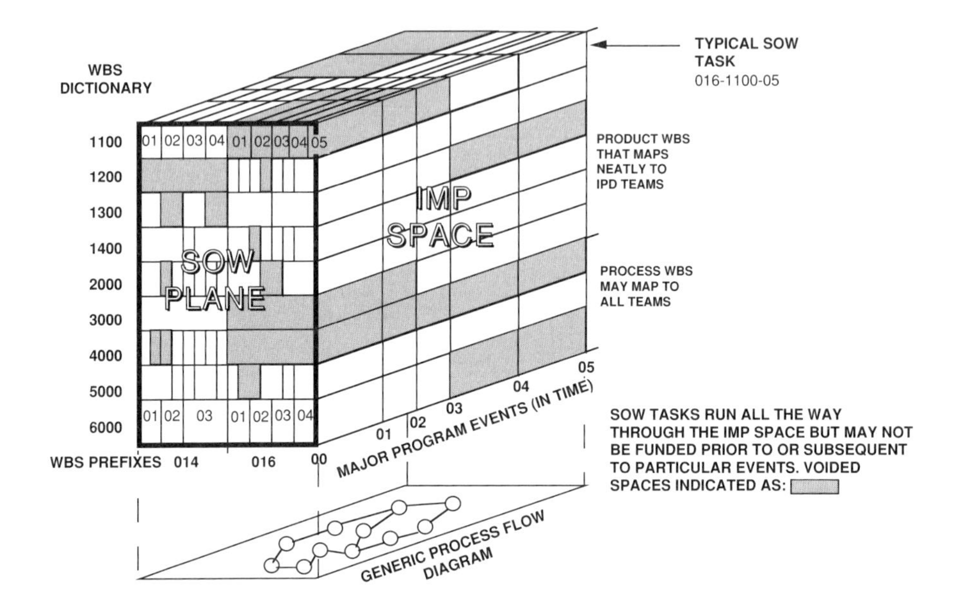

Figure 6-9 IMP space partitioning by event.

management, determines that 20 January 1994 would be the ideal time at which to hold the CDR.

This is only one planning string. It may require thousands of these branching strings to complete the whole planning database for a large program. It quickly becomes obvious that computer technology could be helpful to just capture and organize all of this information. At the time this book was being written, several of the computer tool companies with requirements database tools were beginning to see a potential market for the application of these tools to programmatic as well as product requirements development.

All of these documents have a similar structure of a paragraph, WBS, SOW, or IMP task number; title; and text. These tools could be used to capture the specifications, WBS Dictionary, SOW, and plans and include traceability across the current valleys between them as well as generate the documents, or parts thereof, on demand.

When you have linked up all of the work in strings like the one illustrated in Table 6-3, using manual or computerized methods, you should have a mutually consistent network of management data that provides implementation guidance for people doing the work and management test points for effective program health monitoring.

6.5.4.3 Final IMS development

As we are developing the IMP content we do not obligate ourselves to define scheduling constraints, only to identify the work that must be done. Once the work is defined, scheduling must arrange the tasks corresponding to the significant accomplishments in time, constrained by the events previously located in time. In the process, we will find that some things cannot be

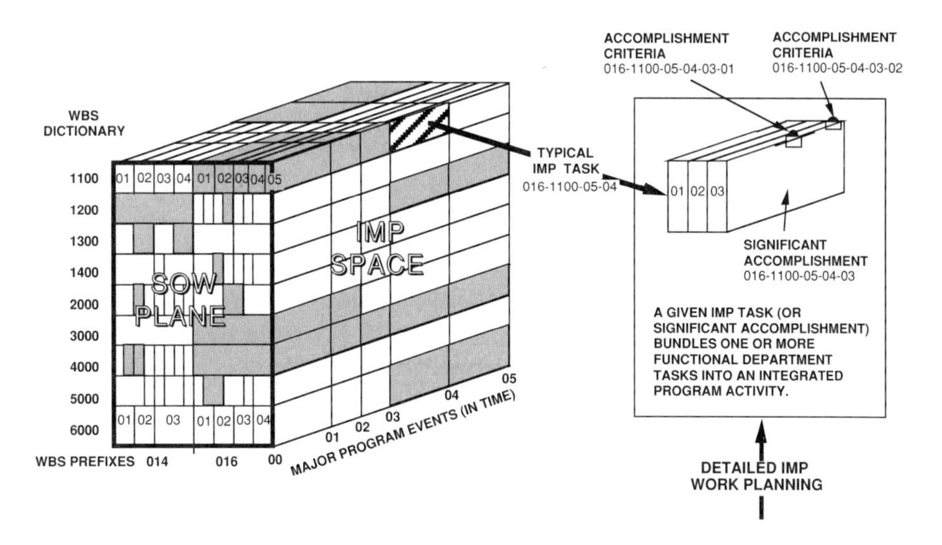

Figure 6-10 IMP task definition.

accomplished as planned in the available time, requiring adjustments in event timing and span times between events. This may require several iterations before all of the inconsistencies shake out and the complete set of planning data is ready for use.

6.5.4.4 Planning process summary

Figure 6-11 combines all of the steps described over the past few pages into a single diagram to better illustrate the intended continuity between the steps outlined. Our approach has been to associate generic functional department work descriptions with WBS numbers defined for the particular product system derived from an understanding of the needed system functionality.

These functional tasks were then combined by the responsible teams into coherent SOW tasks oriented about the product WBS structure. We next determined some major milestones to use in managing program progress

Table 6-3 Single planning string example.

DOCUMENT	REFERENCE	PLANNING CONTENT
SYSTEM SPEC	3.7.1.2	Launch Vehicle Upper Stage
WBS	014-1200	Launch Vehicle Upper Stage Development
SOW TASK	014-1200-01	Accomplish requirements analysis needed to develop the upper stage
IMP EVENT	014-1200-01-06	Item Critical Design Review
SIGNIFICANT ACCOMPLISHMENT	014-1200-01-06-01	Design complete
ACCOMPLISHMENT CRITERIA	014-1200-01-06-01-01	95 % of drawings released
IMS EVENT/DATE	014-1200-01-06	20 January 1994

and projected these events into a space created by expanding the SOW plane into the time dimension. Finally, we further organized the work content of each IMP task defined by a SOW task and a terminal event into a series of specific significant accomplishments each with pre-defined accomplishment criteria.

6.5.5 Contract data requirements list

The contract data requirements list (CDRL) identifies every item of data the customer expects to have formally delivered. This should be driven by the work that will be accomplished as defined by the SOW. Every task listed in the SOW that will produce a data item of value to the customer in managing the program or understanding the technical or administrative flow of events, should be referenced there in terms of its CDRL identification and all of these items collected to form the CDRL. The CDRL items should be determined from the SOW tasks and, once identified, should be fed back into the SOW for reference.

Customers sometimes mindlessly boilerplate the CDRL as they do the SOW rather than deciding in an organized fashion what, of the information that the contractor must produce to develop the product and manage the development, they need for the same purposes. Customers realize that formally delivered data costs contractors more than data generated only for internal consumption so there is a naturally limiting cost boundary at work. The customer should be able to acquire data developed on the program anyway whether the item is on the CDRL or not. This data follows the contractor's format and content definition rather than the customer's, as in a CDRL, and requires a special request on the customer's part to obtain it.

The contractor should maintain an organized list of all program data and make it accessible to all contractor personnel in a program library. A DoD customer calls this list (exclusive of CDRL items) a data accession list (DAL). This list names all of the data the customer can acquire by specifically requesting it and paying a separate fee under the terms of the contract. The contractor benefits by having this same data available for internal use rather than it becoming lost in desk drawers and waste baskets.

The CDRL must include schedule and financial reporting data that allows the customer to determine the health of the program. In addition, the customer should wish to receive technical performance measurement (TPM) data that offers insight into how the technical development is progressing in terms of meeting a small number of key system requirements. Together these three items inform the customer about evolving program cost, program schedule, and product performance.

The customer will normally wish to make the system specification, WBS, SOW, CDRL, and IMP contractual documents requiring a formal process (contract or engineering change proposal) to change. The IMS, however, should be a CDRL item because it will have to change as a function of showing progress, if for no other reason. All other CDRL items should be

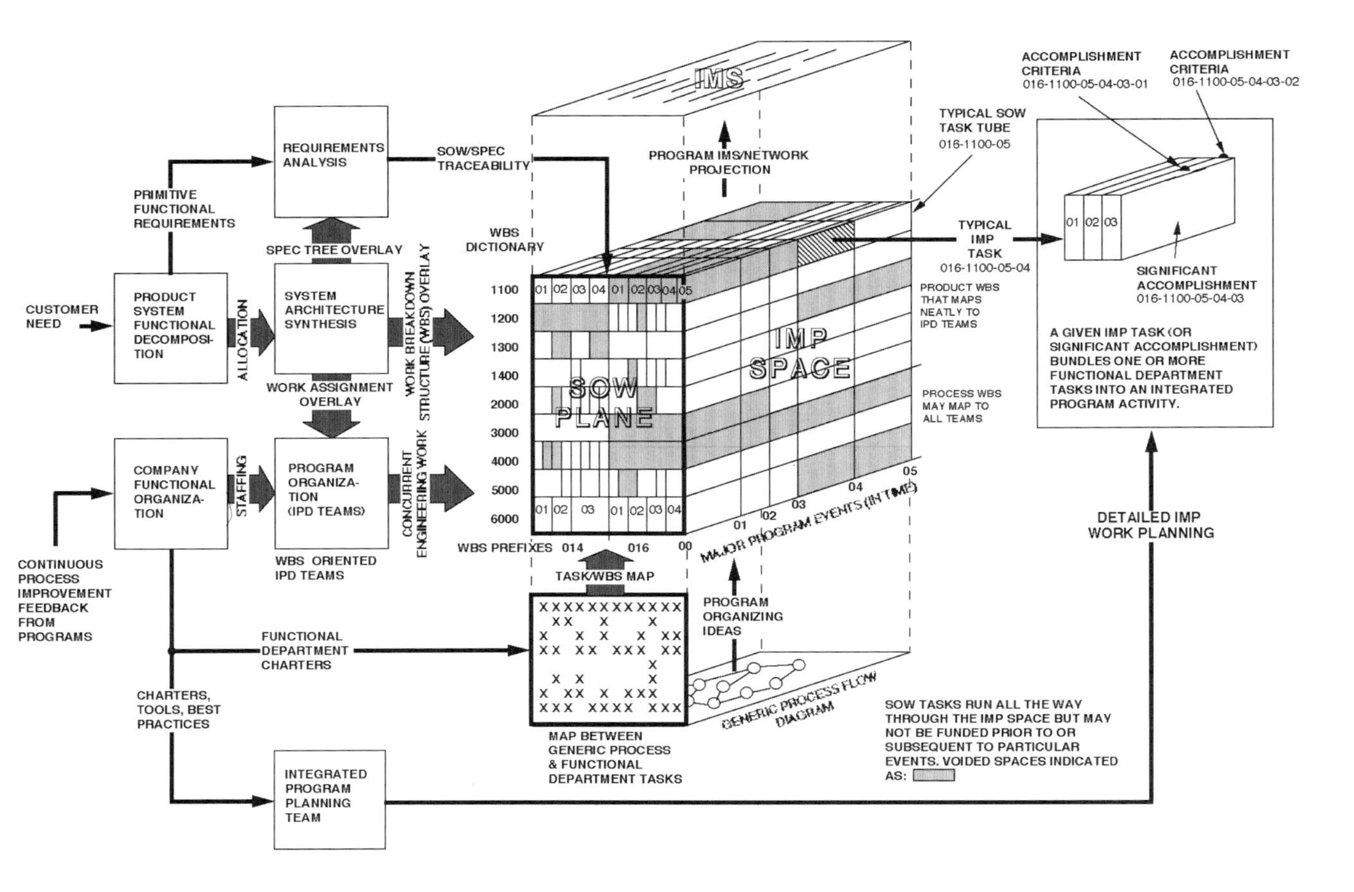

Figure 6-11 Overall program planning scenario

selected with care from the work identified in the SOW to provide the customer insight, at reasonable cost, into the evolving product requirements and design synthesis and into the product development, testing, and production processes.

Ideally, the DAL and CDRL items should be in electronic media stored on a computer network server and accessible from all work stations in the contractor's facility. The contractor can then also quickly respond to a customer request to electronically deliver any item listed in the DAL and in the program library. Very little data is produced today by means not involving a computer. Therefore, most items that a customer would be interested in receiving as a CDRL exists in electronic media within the contractor's business. Yet many customers continue to insist on delivery of tons of paper documents that will fill ever expanding storage space. Alert customers will begin to require delivery of CDRL items in electronic media. This will change the nature of CDRL delivery. Instead of periodic delivery of paper copies of updated documents, the contractor will be required to refresh a customer database at a particular interval or at particular milestones.

The DAL should exist as an electronic data delivery conduit from the program on-line library to a customer. A customer could be charged a periodic rate plus a fee for each call to gain access to this library section. There are some fine precedents for this kind of service in the form of on-line databases open to public access like Compuserve. This kind of automated process would result in important savings for large customers like the government. This arrangement will require special provisions in the contract and good discipline on the part of customer personnel to avoid over-running cost targets.

The company that is able to integrate all of this information product into a central program library has a distinct advantage in concurrent engineering because concurrent engineering is, or at least requires, effective communication of ideas. The people working on the project will also never have to worry about using out-of-date paper copies of documents. And, incidentally, the facilities people in both customer and contractor ranks will even benefit from fewer file cabinets.

6.6 Work responsibility

The IMP must relate program work to work execution responsibility. First, we must decide how we will organize to execute the program. In Chapter 3 we discussed the organizational structure preferred by the author, namely matrix management characterized by: (1) programmatic integrated PDT and (2) functional departments that provide qualified people, proven tools, and procedures to programs. Of course, other forms of organizational structures can successfully execute a program. For small companies with little product line differentiation, forms other than matrix may even be an advantage. Earlier we saw that these alternative structures are actually limiting cases of the general purpose matrix we illustrated in Figure 3-2. This book attempts to focus on a business situation where a matrix is advantageous for the reasons suggested in Chapter 3.

If we are to follow this pattern, our program work must not be assigned to functional department directors, managers, and chiefs, rather to responsible PDT. In the process, if we are not careful, we may develop a team structure that is completely unworkable in execution. The team structure must be aligned with the WBS (product architecture) in order to preclude conflicts in assignment of budgets and tasking to teams.

The product teams must be aligned with the product structure reflected in the WBS because the budget will be aligned with the WBS in order to satisfy the cost/schedule control system (C/SCS) criteria required by a DoD customer or an equivalent entity by a non-DoD customer. In 1991 the criteria for DoD contracts was relocated from DoD Instruction 7000.2 to newly released DoD Instruction 5000.2 but remained unchanged in content. As discussed in Chapter 3, the C/SCS criteria requires us to manage budget through intersections in a matrix of WBS and functional organizations.

We wish the teams to align with the WBS so the budget for all team tasks can be simply assigned. The larger the number of crossovers that exist between WBS, teams, and tasks, the more complex the program will be to manage. At the same time, the WBS must align with the functional architecture allocated from needed system functionality. This means that the WBS must track the evolving functionally derived architecture reflected in the system specification. This combination will result in the fewest possible crossovers between organizational interfaces and product interfaces, referred to as cross-organizational interfaces in Chapter 12, which lead to program problems.

This combination also results in the simplest possible integration task. System integration is a difficult task no matter how expansive. We can do a better job at it if there is less of it to do. Coordination of the organization of the work, product, and performing organization leads in this direction. Figure 6-12 illustrates this point by showing a perfect correlation between a product N-square diagram and an organizational N-square diagram. An N-square diagram displays the relationships between N items by noting these relationships in the square matrix intersections. The N items are listed down the diagonal.

On the product N-square diagram we see a requirement for an interface between two specific items identified on the diagonal. On the organization N-square diagram we see a need for the two teams corresponding to the product items joined by this interface to communicate about this interface. If our PDTs are organized about the same structure that the product system uses, the team communication patterns will match perfectly the product item cross-organizational interfaces we will discuss in Chapter 12. It is precisely these cross-organizational interfaces that traditionally lead to development problems. If there is a complex relationship between product development responsibility and product composition, there will be interface development problems leading to unnecessary cost and schedule impacts.

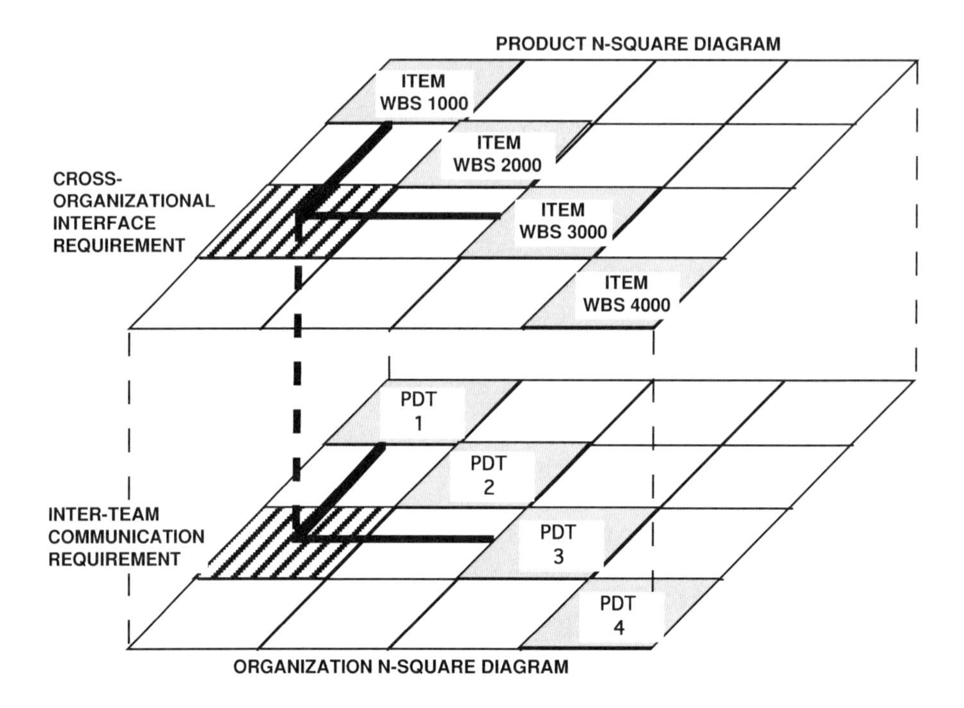

Figure 6-12 Product and teaming N-square alignment.

Either the SOW or the IMP can cross-reference the work to the program organization team structure. Since the SOW lists all tasks, you might conclude that it would be a better place to locate the task responsibility matrix than the IMP. The IMP selectively expands on SOW content linking work to specific events and accomplishments corresponding to those events for management purposes. The author encourages that this matrix be placed in the IMP because the SOW tends to be a simple task list for people with green eye shades while the IMP is a plan for us humans. As the product and team structures mature, we must check for emergence of misalignments between the WBS and the team structure and work toward nulling them out.

Every task identified in the SOW should have some kind of procedural coverage that tells how that task will be performed. This may be in the form of a customer standard (such as MIL-STD-490A), internal company procedures or commercial standards (such as the ANSI series). This can easily be defined in a task/procedures matrix or integrated with responsibility assignment into a SOW task/responsibility/procedures matrix such as the fragment illustrated in Table 6-4.

6.7 *Who plans the program?*

We have left one thing unanswered up to this point. As a result of our wonderful planning work, we know what must be done, who shall do it, when it shall be done, and how it shall be done. But, who should have

Table 6-4 Task/responsibility/procedure matrix.

WBS TASK	RESPONSIBLE TEAM 1	2	3	4	5	6	7	PROCEDURAL REFERENCE
1000.01	X							MIL-STD-1422 (Tailored)
1000.02		X						Company Practice 128.23E
1000.03			X					MIL-STD-490A
1000.04	X							Company Practice 153.5
1000.05					X			Program Manual 14.24B

accomplished this planning work in the first place. Should we let the functional departments engage in bottom-up, grass roots planning integrated by the program? Should the program or proposal team do all of the planning?

The author believes that few proposal teams or program staffs at the time this book was written accomplished their program planning activities in a purposeful integrated fashion, including defining traceability throughout the product and process requirements stream as suggested in Figure 6-2. If the systems approach (or concurrent development) method is useful in developing product systems, why should we not apply it to the program planning process? We, the program team, are after all a system. Let's try it.

First, who should participate? Table 6-5 lists the planning documents discussed above and correlates them with some generic principal functional organization responsibilities. Your company may be organized differently, but these organizational entities are probably fairly widely recognized. While these documents should be developed in approximately the order listed (from top to bottom), we should not permit autonomous work on any of them. There will be insufficient time to sequentially develop these documents during a proposal preparation period of 30 to 60 days. More importantly, their content must be mutually consistent. Figure 6-13 offers a rough schedule showing the relative timing suggested.

The recommended integrated planning approach involves forming a program planning team with membership by representatives from each functional department noted in Table 6-5. Someone identified by the proposal or program manager should lead the team. All members should be

Table 6-5 Program integrated planning team responsibilities.

DOCUMENT	FUNCTIONAL RESPONSIBILITY
SYSTEM SPECIFICATION	Systems Engineering
WORK BREAKDOWN STRUCTURE	Systems Engineering & Finance
STATEMENT OF WORK	Systems Engineering & PDT Managers or candidates
INTEGRATED MASTER PLAN	Systems Engineering
INTEGRATED MASTER SCHEDULE	Scheduling
CONTRACT DATA RQMT LIST	Data Management

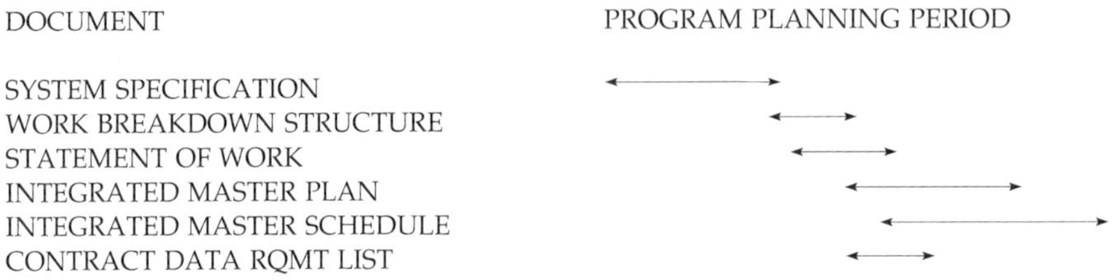

DOCUMENT	PROGRAM PLANNING PERIOD
SYSTEM SPECIFICATION	←——→
WORK BREAKDOWN STRUCTURE	←——→
STATEMENT OF WORK	←——→
INTEGRATED MASTER PLAN	←——→
INTEGRATED MASTER SCHEDULE	←——→
CONTRACT DATA RQMT LIST	←——→

Figure 6-13 Program planning timeline.

physically collocated in close proximity encouraging easy conversation and a close working relationship. They must have available to them good telephone and computer data communications facilities as well as adequate wall space for posting information for integrated viewing.

The suggested integrated approach is not rocket science any more than is the concurrent or integrated product development approach. It simply requires clear definition of responsibilities, cooperation between the parties, excellent interpersonal and communication skills, and a shared appreciation for the discipline of traceability. The system architecture identified in the system specification must be respected in the WBS. All work must be listed in the SOW and linked to the WBS. All events, accomplishments, and criteria in the IMP must link to the tasks defined in the SOW. All tasks appearing on the IMS must correspond with the SOW and IMP tasks and all events on the IMS must correspond to those respected in the IMP.

Throughout the planning process, the selected PDT Managers must contribute to WBS and SOW development and maintain vigilance for crossovers between WBS, team definition, and work responsibilities. They must also work to develop a cost estimate in a proposal situation. Also in a proposal situation, this team will have to coordinate their work with the management volume writing team.

This planning activity probably cannot be accomplished in a straight line fashion, While preparing the IMP we will get insights into changes in the SOW that may ripple through other documents. All team members must have available to them the full content of all of the evolving documents and be familiar with that content through an almost continuous conversation between team members and access to the actual text on their computer screen and a stickup on a wall in close proximity.

Team members must respect the hierarchy in Figure 6-2 and the content of the documents should be developed in the sequence illustrated there. Each component of each document expands into lower tier document details. This pattern repeats through the hierarchy. As this expansion is developed, the traceability links should be captured. The information developed by the team members assigned to each document must be constantly checked for traceability and consistency by someone responsible for planning integration.

Some readers may think that the material in paragraph 6.5 should have been placed prior to the planning material in paragraphs 6.3 and 6.4. At the same time you have to be familiar with the Air Force initiative before appreciating the opportunities to accomplish the planning work in an integrated fashion. The author agonized over this dilemma for some time until concluding that the reader should first understand the planning relationships outlined in the Air Force initiative, then be exposed to the integrated planning team concept not explicitly included in the initiative. You should now scan the previous material in this chapter with the perspective of the integrated planning team concept in mind.

6.8 A generic SEM/SEMP for your company

Some people greeted the workup to release of MIL-STD-499B with screams of, "The sky is falling." You would think that, after many years of contract performance under its predecessor 499A, all DoD contractors would have put in place the fine system engineering activities that they had been describing in the many SEMPs that they had written over the years and submitted with proposals. The SEMPs were never contractual and, sadly, often were never opened by the contractor subsequent to the contract win. So, the adverse reaction to 499B was often based on not now having in place, despite the fine stories told in SEMPs past, an effective systems approach and a concern for how to acquire an effective process in a reasonable time. Many companies dispatched people to the several MIL-STD-499B classes offered only to have them return with confirmation that they were in very big trouble.

Whether your company must respond to MIL-STD-499B, some other customer standard, or has no constraints on your process, you really should have in place an effective systems approach because it will result in a product with better value for your customer and more business for your company. If your competitors are able to put in place an effective systems approach and you do not, you will have great difficulty matching their product cost, schedule, and quality.

The book *System Engineering Planning and Enterprise Identity* includes a copy of a complete generic integrated System Engineering Manual/System Engineering Management Plan (SEM/SEMP). This document can be used as a basis for your company SEM/SEMP. It is possible to purchase this document on a computer disk such that you can edit it directly for your needs. It is also possible to purchase a computer disk for the process diagrams and task database included within the appended SEM/SEMP. You may very well also have employees or managers who are fully qualified to write such a document from scratch and only lack the feeling of urgency that you need an internal process definition. It is clear to the author, at least, that every enterprise should have a written process description, actually follow that process description, and incrementally improve that process over time based on lessons learned.

It does not follow, of course, that by acquiring a good SEM/SEMP that you will overnight become an excellent systems house, though the process of writing one can be quite an education in the right direction. This document must be matched to a current reality within your company. You will need to monitor your performance to this standard and provide the machinery to force compliance with your own standards. The techniques discussed in this book and elsewhere on continuous process improvement should be applied to your system engineering process to uncover weakness and to understand useful priorities for their correction.

You will also need a way to educate your work force in your system engineering process. This can be done at low cost through brownbag sessions in your plant for your motivated employees which happen to be the very ones you wish to retain. The SEM/SEMP can be used as the text book for these sessions. You should also cooperate with your local National Council on Systems Engineering (NCOSE) chapter and a local college or university to establish or support a system engineering certificate program that qualifies for your company's tuition reimbursement program.

Given that you have a written procedure for performing system engineering, a way to educate/train your work force in performing that procedure, a means to continuously improve upon that procedure, and a way to encourage compliance with your process, you are on the road to success in performing system engineering effectively. You will also have little to fear from the standards any customer may impose upon you.

chapter seven

Specialty integration & concurrent development

7.1 The past

The term specialty integration was popularized by MIL-STD-499 and 499A. One of the three major sections of a system engineering management plan (SEMP) described in that standard had this title. Specialty engineering disciplines arose as it became obvious that the design engineer could not master the creative process of design synthesis and the ever expanding number of specialized design views. As these specialized views, like reliability and maintainability, became accepted, it soon became obvious that someone had to integrate the work of these many people and the system engineering process was born.

Specialty engineering disciplines developed effective methods to define their requirements, communicate them to designers, and assess the design for compliance. Each specialty engineering discipline became a separate house focused narrowly on its own agenda. Societies came into being encouraging this inward focus. Functional organizations were founded for each discipline encouraged by DoD customer organizations in these same fields. DoD prepared military standards and handbooks on how to perform these disciplines and required their application and compliance on contracts.

Something went wrong in the evolution of the systems approach that resulted in the specialty disciplines acquiring the habit of serial work performance with physical collocation by functional organization rather than product team. Designers would create designs and the many specialty engineering disciplines would independently review the designs for compliance with their requirements. This pattern of work is called by many transom engineering. The designer throws the drawings over the transom to an isolated specialty engineer who notes problems that must be corrected and throws the drawings back. A tremendous amount of time is wasted in this process that could be avoided by a simple conversation between the designer and the specialty engineer before the designer committed to a particular design feature. The transom engineering analogy is very apt in illustrating the antithesis of what the system engineering process was supposed to be.

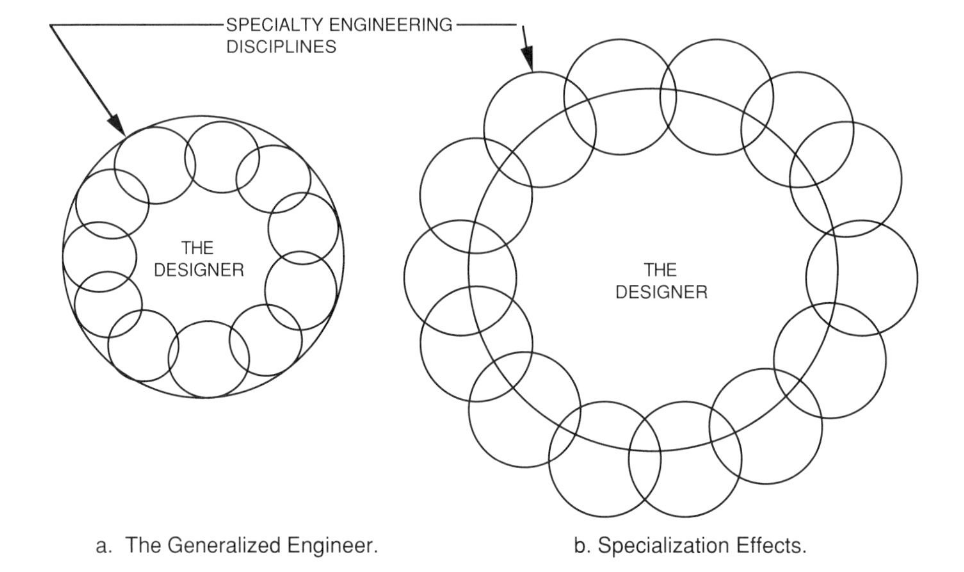

Figure 7-1 A changing world.

Figure 7-1 illustrates the specialty engineering problem from the designer's perspective. The author was first exposed to this view by Mr. Steve Landry, at the time a system engineer from General Dynamics Convair Division, at the GD Systems Engineering Seminar. Many years ago one designer could master everything needed to synthesize the requirements into an effective design solution. Times were simpler and so were systems. The designer could master what was known about reliability and safety. If the product embraced these characteristics it was because the designer put them there. There are some old designers still around who lived in this world and resent to this day the specialty engineering crowd trying to influence what they see as their prerogatives.

Today, the designer simply cannot master his/her engineering discipline and all of the specialty fields as well. This is the very reason for system engineering and structured decomposition of large problems into related smaller ones. The challenge is to provide an environment within the work place such that the designer and the crowd of specialists become one again. The stage is now nearly set for the entry of the concurrent or simultaneous engineering initiative and a return to what the founding fathers of the system engineering process had in mind.

Those programs that have been well executed from a specialty engineering perspective and produced the required product within cost and schedule constraints have very fortunate customers. Let us picture for a moment that the customer for Program XYZ has prescribed through the program SOW and CDRL that there shall be twenty specialty engineering disciplines on the program. Since the reader will immediately say that this is an outrageous number, let us list the twenty the customer has pre-determined in their contract language. Do any of these seem outrageous? No, most of them are fairly common on many programs. It is not hard to arrive at twenty specialty engineering disciplines on a program.

reliability	human factors	logistics support analysis (LSA)
maintainability	parts engineering	survivability & vulnerability
availability	materials and processes	transportability
mass properties	system safety	Life Cycle Cost
aerodynamics	system security	quality engineering
structural dynamics	thermodynamics	producibility
guidance analysis	contamination control	

In most engineering organizations the older disciplines requiring use of mathematical modeling as well as very technical engineering knowledge and skills (aerodynamics, thermodynamics, etc.) would be grouped into a system analysis department. These are the disciplines that the design community has more or less accepted. The newer disciplines that do not normally require design skills like reliability, maintainability, and safety, would commonly be grouped in a system engineering functional department. These disciplines have yet to prove themselves to the design community and practitioners of these disciplines have to fight every day for their self-respect

But, let us accept that it is a perfect world that Program XYZ is functioning within. Each specialty has the following approach to their activity:

a. define specialty requirements
b. **communicate specialty requirements to the designer**
c. **ensure the designer understands the requirements**
d. **offer examples of how these requirements have been satisfied on similar systems**
e. review the engineering drawings for compliance with specialty requirements
f. **instruct the designer on failures to comply**
g. re-evaluate changes for compliance

Each of the process steps in bold requires communication with the designer. There are 20 specialists each of which must communicate a minimum of four times for an average of one hour each with each designer. This is 80 interactions and 160 program man-hours for each designer. Expand this number by the number of designers simultaneously working the program, say 40. We now find the specialty engineering designer contact event count is up to 3200 ($20 \times 4 \times 40$) on the program through CDR requiring 6400 man-hours.

Let us say that each specialty engineering discipline is especially effective in ensuring its requirements are clearly understood by the designer. Each discipline has prepared a checklist and they will expect the designer to read and sign the checklist signifying his/her understanding. These checklists commonly represent a digestion of the content of a military standard called out in the SOW or system specification. Let us say that the average checklist has 50 items. The designer will have to read and understand 1000 checklist items (50 items/checklist times 20 checklists) in items b and c above in addition to the other requirements in the item specification. The total program checklist event count becomes 40,000 (40 principal engineers times 50 items/discipline times 20 disciplines). Heaven help the designers if all of the specialty engineering disciplines were to become uniformly effective in influencing the design process in this environment.

In what became the traditional specialty engineering approach, each specialty engineer would talk to the designer independently yielding no opportunity for interaction and synergism between specialty engineers. The only one who was exposed to all of the conversations (given that all specialty engineers pursued their responsibility with equal vigor) was the designer. If only one person on a program is to have this experience, it should be the designer, but why must only one person have this experience? True concurrent engineering implemented with a DIG, war room, physical collocation, and good leadership will permit an explosive simultaneous conversation with everyone's information available to all.

Clearly, our history has shown that products that comply with specialty requirements provide better value than those that do not. The challenge is to provide a teamwork environment that frees the designer to apply his/her creative genius to synthesizing the complete set of requirements within the safety of boundaries protected by the specialty engineers. This environment must ensure that all needed specialty views are energetically pursued yet ensure that the designer has adequate time to think and to do the creative design work. We need an orderly environment as far as the specialty community is concerned. But, the designer needs a creative environment with an absolute minimum of constraints. Somewhere there must be a balance point between these conflicting needs. That is our challenge.

7.2 The future

Before we approach a practical solution that can be applied today, let use first consider what might be the ultimate solution to this dilemma. The author is convinced, after seeing some of the work going on at General Dynamics Space Systems Division (similar to work at other firms no doubt) in the area of RAMCAD, that the ultimate solution to specialty engineering integration will take the form of design rules embedded within the CAD tools that protect the designer from violating the constraints now expressed in checklist form.

The designer need not read and understand any of these checklists (though the designer might be more efficient the more he/she understood of those requirements). If the designer of a circuit board places two resistors too close together such that they are exposed to an excessive thermal condition. the CAD tools will not permit the error to be made. A signal will flash on the screen noting the error and what to do about it. The design would never have to be reviewed by the reliability or thermal experts, or the review could exclude certain tedious elements. Specialty engineers in this environment ply their trade by helping to program the CAD tools for error avoidance.

Some of these kinds of features are being built into CAD tools today. It is likely that this environment will be very effective for some specialty engineering disciplines and not so effective for others as a function of the ease or difficulty in translating the specialty discipline requirements into CAD rules. Many aspects of system safety and human factors engineering will continue to demand human judgment for a very long time, perhaps forever. At the same time, some human factors elements like the size of access openings and accessibility for adjustments can be very easily introduced into 3D CAD tools.

These demonstrations of future capabilities are very exciting to behold, but generally are not available now for deployment into effective tools for use on programs. In the near term, until the ultimate solution is available, we must find another solution to effective specialty engineering integration.

7.3 Concurrent development for the present

PDT activity at any one moment is a function of program phase, product item characteristics and applicable technology (software versus hardware, for example), and team position on the development progression. Normally, PDTs will not be assigned until the product architecture has been established by early program work accomplished by the PIT. A project in its earliest phases should begin with the PIT only growing out of a proposal or marketing experience. This team should accomplish early program work and accomplish the planning work for later program phases including defining teams corresponding to major product system elements. Once PDTs are identified and staffed, they should follow a predictable progression outlined in subordinate paragraphs.

7.3.1 Concurrent development and PDT overview

Throughout the period characterized by the term specialty engineering integration, the emphasis was on the specialty disciplines imposing their requirements on the product designer. The logistics engineer was supposed to influence the product design to be responsive to support concepts. The manufacturing engineer had to ensure that the product design included producibility features. Similarly, other disciplines attempted to influence the product design in a unidirectional fashion. The author recognizes the concurrent development approach as an expression of the originally intended system engineering process with one exception. That distinction is that concurrent development encourages a bidirectional development conversation between the product design engineer and those responsible for the design of the production process, the logistics support concept, and the operational deployment and employment plans.

The emphasis today is on all of these designs being developed together, each influencing the others to arrive at a true system optimized solution, not just an optimized product design solution. Just as a manufacturing engineer may ask a product designer to replace a forged beam with a larger cast beam to simplify the manufacturing process, it is just as valid for the product designer to ask the manufacturing engineer to use a forged beam rather than a cast one to preclude a beam size that adversely influences the space available for needed instruments. This should not be a one-way conversation only addressing the design of the product. The PDT responsible for a product element should also be responsible for the corresponding factory, logistics, and operations designs wherever possible. The team must develop a unified solution to the overall problem, not just a product design that respects specialty requirements.

This significant difference of a bidirectional rather than unidirectional conversation between the product designers and specialty engineers has the effect of encouraging a better degree of timely teamwork between the team members whereas the old view of the process encouraged shell-shocked product designers overwhelmed by checklists.

Figure 7-2 illustrates an overall view of the PDT-implemented concurrent development process. It is very difficult to illustrate the linear relationship of tasks and the explosive vertical dimension of the decomposition and integration process comprising the structured development process. The spiral diagram included in *A Spiral Model of Software Development*, Tutorial: Software Engineering Project Management, IEEE Computer Society Press, 1988 by B. W. Boehm; and the vee diagram encouraged by Kevin Forsberg

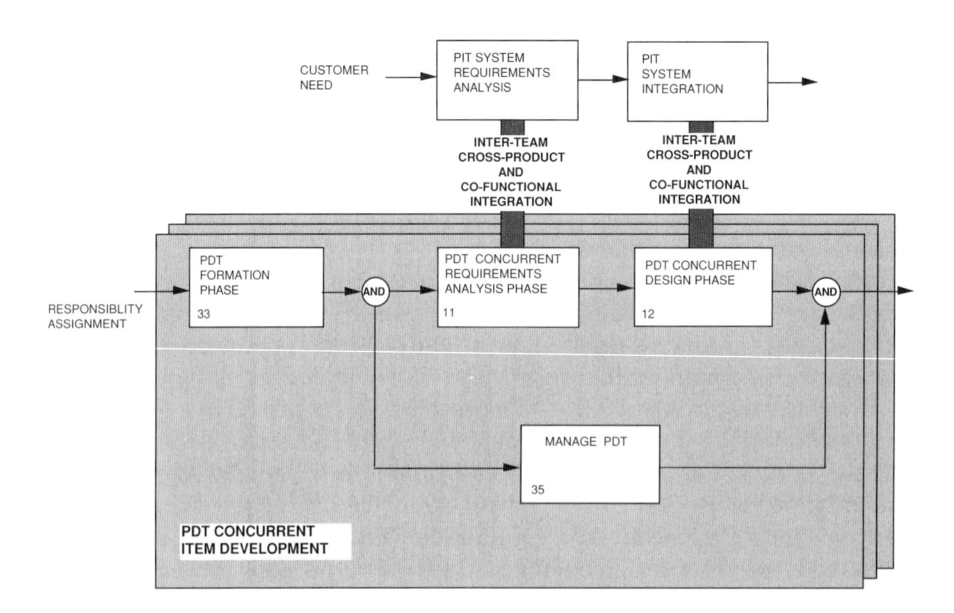

Figure 7-2 Overall concurrent development process.

and Harold Mooz in *The Relationship of System Engineering to the Project Cycle*, 1991 NCOSE Symposium Proceedings are attempts to solve this difficulty. Both of these diagrammatic treatments are effective, but the author has chosen to use the simple relationship depicted in Figure 7-2.

You will note in Figure 7-2, that there is one PIT process but many PDT processes, one for each product team recognized by the program. Further, a PDT may find it necessary to establish sub teams and principal engineers below that level. The PIT is responsible for developing the system architecture and assigning the responsibility for development of those architecture elements to PDTs. These teams come into being through this process. The PIT must concurrently develop the system requirements and monitor the lower tier development of requirements within the PDTs for traceability to the system requirements. As the PDTs complete their requirements analysis process, the PIT must review and approve those requirements and authorize the PDTs to initiate the design phase. During the design process, the PIT must monitor the PDT concurrent design process. In so doing the principle PIT targets are detection of sub optimization and interface problems between items for which the different PDTs are responsible.

7.3.2 The concurrent engineering bond

Product requirements analysis, concept development, preliminary design, and detailed design work must be accomplished in concert with a host of specialty engineering disciplines. Figure 7-2 illustrates this activity in the form of a concurrent engineering bond depicted as a dark band joining tasks that must be performed in a concurrent fashion involving intense cooperation between personnel responsible for the tasks thus joined. You will note that the integration work taking place between PIT and PDTs on Figure 7-2 involves primarily inter-team cross-product and co-functional integration. The concurrent engineering bond within an PDT involves primarily intra-team co-product and cross-functional integration. It was the recognition of these different views of the integration process expressed in this diagram that eventually led the author to write this book

This bond signifies the shared experience and knowledge of the team members. Our intent is to create a work environment within which many individuals can cooperate toward common goals. This cooperation requires communication, the keystone of concurrent development and the medium through which the team members share and thus become one.

The concurrent development process does not encourage concurrent definition of requirements and design development, a common mistake that results in a leap to point design solutions later supported by requirements that are satisfied by the design solution. Requirements definition and design solution must each be accomplished concurrently but there should be a sequential relationship between requirements definition and design work with requirements work leading.

7.3.3 Team formation

The PDT comes into being as a result of a PIT decision in a process that will be more fully explained in Chapters 11 and 12. Two team staffing alternatives are offered that the program and company management staffs must decide upon: (1) team leadership selection and (2) imposed leadership. Some integrated or concurrent product development authorities have encouraged that the team should pick its own leader. That may be the best approach, but program and company management are going to have to decide if that is within their range of acceptability. It is only suggested here that the program manager, to whom this person will report, should have something to say about the selection. Otherwise, the team may flounder for some time until it can deal with the leadership issue. A bad choice of an imposed leader, initial or permanent, may, of course, have an even worst influence on the team's prospects for success.

The solution offered in Figure 7-3 is to initially staff the team from available personnel (supplied by the functional departments) and appoint an initial leader who will have the responsibility to complete the team formation process. If this person gains not only program management's confidence but the team's as well, then he/she is probably the right one to lead the team to completion of its work. The last block on the right of the first tier allows an opportunity to change the leadership based on the team's experience in forming.

The team members must reach a common understanding of their goal expressed in the IMP and agree upon the process to be used to attain that goal. They must all understand the communication aids available (computer networks, bulletin board, telephones, specifications, the IMP/IMS, and meeting area tools) and how they will use them. Most of us humans are wonderfully

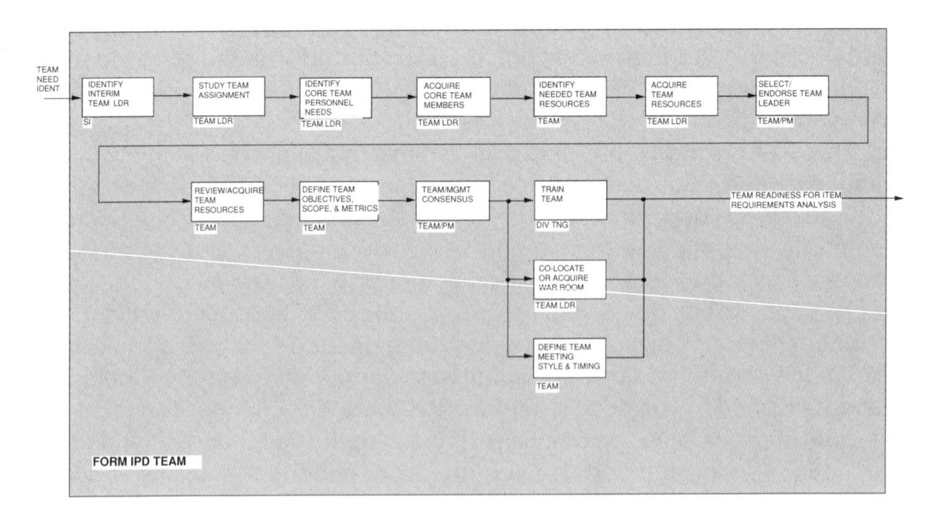

Figure 7-3 Team formation process.

effective goal-striving creatures. We can together satisfy very challenging goals if we understand and agree with them, are well led, and have the necessary resources. Good leadership will have the effect of infusing us with a common purpose so strongly felt that we must attain it. The concurrent bond reflects this intensity of feeling through which we are encouraged to share our specialized knowledge toward our common purpose.

The final step in team preparation is to train the team in concurrent development practices. The nature of this instruction is a function of the current position of the company on its continuous process improvement path to PDT capability. If the company has only recently shifted to this approach, a one week course may be needed. If this is the third program employing PDT, the team may only require a two hour refresher course in group dynamics.

7.3.4 PDT requirements development

Each PDT must implement an organized requirements analysis process to define item requirements derived from the information supplied to them by the PIT. Figure 7-4 illustrates one view of a generic requirements analysis process appropriate to a concurrent development environment employing a structured development process. The team must first study allocated requirements provided by the PIT. Thereafter, a requirements analysis task is included for each specialized discipline. Note that this includes not only product requirements but requirements for the production process, logistics support, quality implementation, and operations activities. All of this requirements work must take place concurrently with a sharing of the resultant information taking advantage of the communications resources described in Chapter 5.

The concurrent development bond suggests this intense conversation leading to a coordinated set of requirements shown flowing from the right side of the blocks. These output arrows are annotated with paragraph numbers for a simple requirements document style guide featured in the book *System Requirements Analysis.* A study of the structures of the specification types identified in MIL-STD-490A, led the author to a conclusion that the structures were very difficult to mate up with a modern system requirements analysis process resulting in the conclusion that a new specification structure would simplify the requirements integration process. Figure 7-4 suggests an alignment between the process and the structure of the data developed by the team.

Every PDT should receive from the PIT the following inputs:

a. An identification of the architecture elements superior to the item for which they are responsible
b. A list of functions allocated to the item
c. The requirements for the immediately superior item in the system architecture

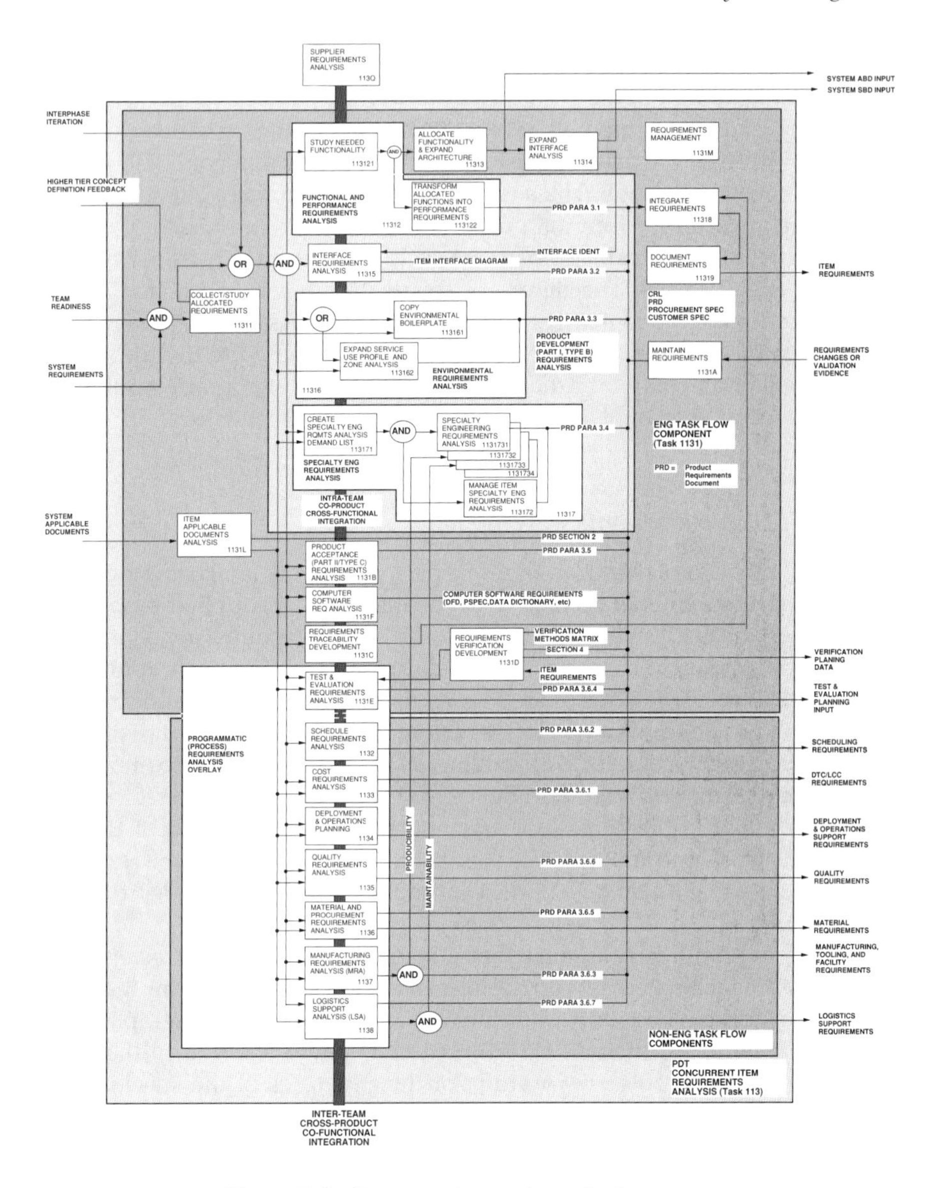

Figure 7-4 Item requirements analysis process.

d. Identification of required external interfaces in list, n-square diagram, or schematic block diagram form
e. System environmental requirements
f. A system use profile with the item mapped to the process steps of the use profile or some other means to help the team understand the environmental requirements imposed on the item
g. A list of allocated specialty engineering constraints
h. A list of required customer applicable documents that must be complied with

During the item requirements analysis process, someone must coordinate and integrate the results. This work could be performed by the team leader or by a system engineer appointed by the team leader depending on the magnitude of the process. The integration work should entail:

a. Identification of conflicts in specialty requirements like high reliability and maintainability figures and snow load on a hot tin roof
b. The item requirements must be checked for traceability to parent requirements
c. Identification of a verification approach for each requirement and coordination of these approaches with analysis and test planning activity
d. Review for unnecessary requirements and exclude same
e. Review for completeness and add other needed requirements

While the team leader or appointed system engineer integrates the requirements within the team (intra-team cross-functional co-product integration), the PIT must monitor the evolving requirements set for the item to assure they are compliant with system requirements. This includes integrating the item verification needs with an integrated system test plan and other system level activities. The PIT reliability person should evaluate the item reliability requirement for compliance with the allocated failure rate, the PIT mass properties engineer checks the item mass properties requirement for compliance with allocated weight, and the PIT life cycle cost analyst checks item compliance with the cost allocation. These are all cases of inter-team cross-product and co-functional integration.

7.3.5 Concurrent design development

Given that the team has developed a set of requirements and a design concept for the team item and those requirements and concepts have been approved by the PIT, the PDT is ready to begin preliminary design leading to sketches, layout drawings, validated requirements, and analysis reports for the product. At the same time the product design engineers are leading the product design effort, a design is concurrently being crafted for each of the other development channels: manufacturing process and facilities, tooling, procurement and material, quality, test, operations, and logistics. The design agents in each of these cases must actively interact to evolve mutually compatible concepts that evolve from and survive frequent concurrent team meetings where each party must expose the features of his/her concept and answer questions from the other design agents and specialty engineers.

Figure 7-5 illustrates this concurrent design process. It is composed of two major steps, a concurrent concept development and selection (or preliminary design) phase and a concurrent design (detailed design) phase. In the first phase, the team members join to synthesize the requirements they developed in the previous requirements analysis phase. Alternative product, operations, logistics, and manufacturing design concepts are developed

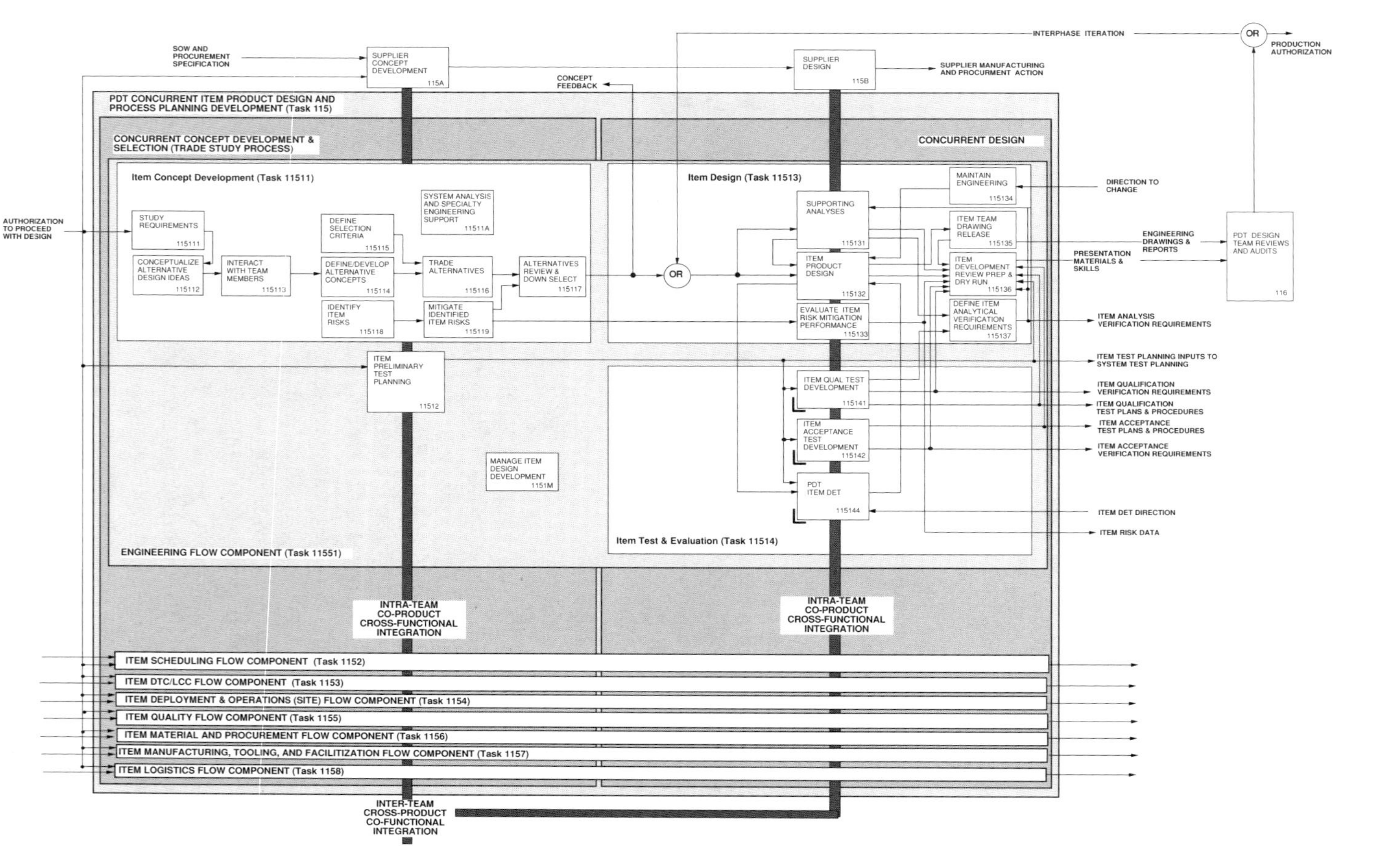

Figure 7-5 Concurrent PDT design process.

creatively by team members and mutually evaluated through a trade study process and the optimum concept set selected. The team moves from preliminary design to detail design after they have demonstrated to the PIT that they have validated the previously approved requirements with a credible design concept for the product and associated manufacturing, logistics, and operations activity. On a DoD program this point occurs at the preliminary design review (PDR).

Throughout the preliminary design phase, the teams use the Interim Common Database (ICDB) information concept to capture development data for definition of the current baseline and communication across the team members and teams. Teams review ICDB information in meeting rooms employing the real-time projection review and update technique and program information directly from the DIG for internal reviews. The whole team is physically collocated and they are able to carry on easy cross-functional conversations within this space. Their work space includes at least one large wall upon which they may place materials they find useful to stimulate a synthesis of their individual team members' ideas.

In the concurrent design phase, the PDT develops the selected design concepts concurrently into a complete set of product engineering drawings, detailed manufacturing and quality inspection planning, and test procedures. This phase is characterized by the most intense product integration activity in the complete development process. This integration activity is covered in Chapters 14 and 15.

7.4 The generic specialty engineering process

A system engineer charged with the responsibility of integrating the work of several specialty engineering disciplines should be familiar with the work these specialists do. Summary descriptions are included below for several disciplines. All or most of these disciplines follow the work pattern presented in paragraphs 7.4.1 through 7.4.4.

All specialty engineering disciplines follow a similar pattern of behavior with respect to design engineers focused on identifying requirements, encouraging understanding of the requirements, concurrent support of the designer while they synthesize requirements into a concept followed by expression of the concept on design drawings, and assessment of the design solution for compliance. This generic process was orinally described by the author in the book *System Requirements Analysis* published by McGraw-Hill in 1993. It is included here with the permission of McGraw-Hill. There is a spectrum of implementation possibilities in this process from focus on compliance assessment to effective concurrent engineering. We will assume here that the organization is capable of concurrent engineering.

The first specialty engineering challenge is to identify specialty requirements for system elements, each of which has a principal engineer or integrated design team leader assigned overall responsibility. Once the requirements are identified, the specialty engineer must help the designer understand the requirements and ways that these requirements have been complied with in the past. During the concept and design development process, the specialty engineer must interact with the designer to concurrently assess compliance while the design unfolds.

Specialty engineers responsible for logistics support, operational employment, and product manufacturing should be simultaneously designing the logistic support system, operational employment process, and manufacturing process (tooling, facilities, manufacturing flow, etc.) while the product design is in progress. As a result, the design engineer functions as a specialty engineer with respect to their activities. Taken together, we see that all of these people must forget about who is a designer and who is a specialty engineer and form a cross-functional team producing a coordinated product design, manufacturing process, and employment process. The goal for the team should be to become the equivalent of one all-knowing engineer.

7.4.1 Concurrent requirements definition

7.4.1.1 Requirements identification responsibility aid

We must have a foolproof way to communicate to the several specialty engineering analysts that they have a specific requirements analysis task to perform. We must have a way to place a clear demand upon them to provide a specific service. One method for doing this is to use a Design Constraints Scoping Matrix (DCSM). The matrix, an example of which is illustrated in Figure 7-6, correlates the different engineering specialties on the left margin with the architecture elements across the top of the matrix. An appropriate symbol in a matrix intersection means that the responsible engineering

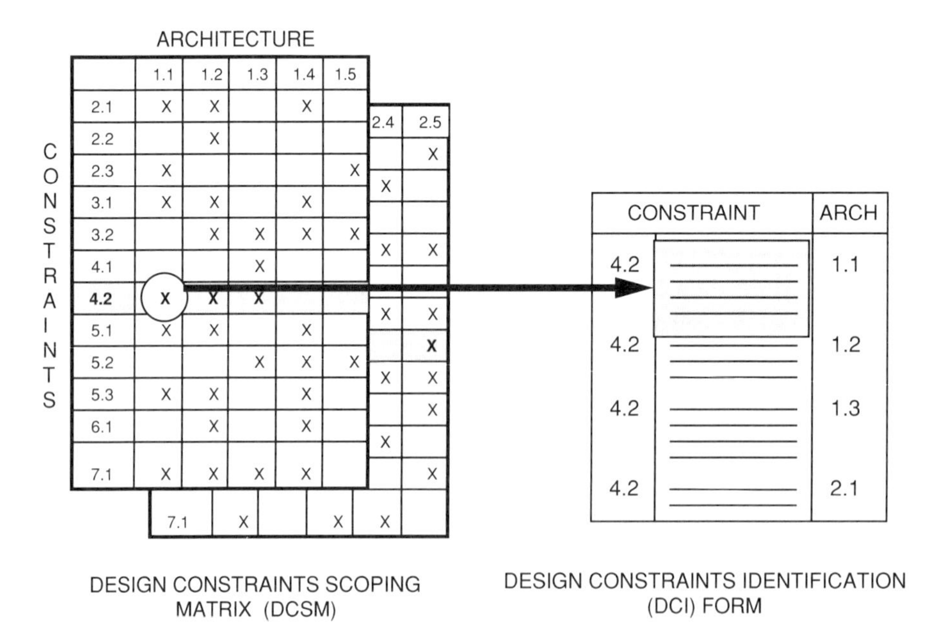

Figure 7-6 Typical design constraints scoping matrix.

specialty discipline (left margin) must identify one or more constraints for that discipline against that architecture element (top margin).

The top, architecture identification border may be structured to permit wide flexibility in form use. The example identifies components using Architecture ID codes to designate system elements, but WBS could be used instead of, or in addition to, these codes. Obviously one matrix for a complete system may become completely unwieldy because of the number of elements in the system. This can easily be handled through multiple pages each with the same vertical axis defining the constraint categories. This also allows us to use this approach in a progressive way as the architecture expands through the application of structured decomposition.

We place a symbol in each intersection to denote whether or not the constraint category applies to that architecture element. We can use a simple "X" to indicate yes and a blank to indicate no. In this case, the specialty engineer responsible for a given specialty constraint category simply responds by writing one or more constraints against each item marked "X" in his/her row.

But, we may also use the matrix to capture a more complex reporting mechanism in addition to tasking the specialty engineering analysts. The system engineering function could first mark up the matrix with "Xs" as indicated above. The matrix (some subset of the total matrix) is then passed around to the specialty groups for review. Each specialty group annotates the matrix to indicate the anticipated impact or difficulty using one of the codes listed below and returns the marked-up matrix to system engineering for integration and database update (this action can be performed more efficiently in a fully automated, on-line fashion, of course).

- X The constraint applies to this element. No difficulty is foreseen in implementing the constraint.
- Y The constraint applies to this element. Additional study or analysis is required to determine the impact of implementation.
- Z The constraint applies to this element. A serious impact is foreseen in implementing the constraint in terms of cost, schedule and/or performance risk.
- — The constraint has been reviewed against this element and found not to apply.

  A blank (no entry) indicates that the analysis related to this intersection has not been completed.

7.4.1.2 Requirements capture

The engineering specialty analyst responds to the final DCSM by applying his/her specialized tools and procedures to identify design constraints for which they are responsible, often by reference to applicable documents appropriate to the contract. These documents must have been screened earlier (preferably in the proposal phase) for any needed tailoring and the tailoring agreed upon with the customer (see Chapter 5). Alternatively, the specialty analyst may appeal to a system mathematical model to allocate a requirement value from parent to child through each level and branch of the architecture. This response could occur within the environment of any of the three basic requirements analysis strategies.

Freestyle approach — In the freestyle strategy, the analyst responsible for a specialty discipline simply ensures that he/she writes one or more constraints against each architecture item defined by the matrix and provides them to the appropriate principal engineers. This could be in the form of a tabular printout of architecture versus requirement values, an applicable document referencing requirement statement that applies to all architecture elements, or a series of statements with blanks that are filled from a tabular listing as a function of architecture.

Cloning approach — In the cloning strategy, the specialty analyst might provide the person responsible for preparing boilerplates with the generic information that applies across all of the elements that might use a particular boilerplate and a tabular list that contains the specific values for each architecture element. Each principal engineer using the boilerplate would simply fill in the value from the table for that item into the blank provided in the boilerplate text.

The flowdown strategy — The flowdown approach is especially appropriate to many specialty engineering disciplines such as reliability and maintainability. It is accomplished by partitioning the requirement value for a parent item into component values for each child item. No matter what strategy is used, flowdown should be used as the detailed mechanism for numerically valued specialty requirements.

The structured strategy — In the structured approach, each specialty engineering analyst would be required to define one or more specific design constraints for each intersection marked with a letter character on each design constraints scoping matrix, as in the other cases. The difference is that they would be required to respond in a definite time frame for given architecture items on a special form provided for that purpose. Figure 7-7 illustrates a typical Design Constraints Identification Form (DCIF). This form could be used directly in a manual implementation of design constraints identification or as a computer data entry form.

7.4.1.3 Constraints integration

The architecture element principal engineer or PDT should be held accountable for integrating the effects of the specialty constraints. The principal engineer will normally be a system engineer at the highest system levels, possibly a PDT leader at intermediate system levels between segment and subsystem (on programs that use teams), or a subsystem or component design engineer at lower system levels. All of these principal engineers will have different degrees of skill and experience with specialty engineering integration and some may need support from system engineers.

There are two alternative applications of a system engineering function, or PIT, in the integration process. First, they could be used to audit the performance of the principal engineers in this process and provide feedback

ANALYST	DATE	LSA REFERENCE		ACTION	
DCC CODE	DCC NAME			REV	DATE
FUNCTIONAL REQUIREMENTS INFORMATION				ALLOCATION INFORMATION	
LINE ITEM	DESIGN CONSTRAINT STATEMENT	SOURCE REFERENCE	CRIT	ARCH ID	STAT

Figure 7-7 Design constraints identification form.

to principal engineers about their conclusions. Secondly, system engineers could be assigned to teams to do the integration work for the principal engineers. The particular program team must evaluate the skills of its personnel and select an appropriate approach for their situation.

In addition, the system engineering community should be tasked with coordinating the specialty engineering requirements analysis activity to include: (1) ensure all required disciplines have access to budget to perform their task, (2) monitor that each required discipline has qualified personnel assigned to specialty requirements tasks, (3) verify that each discipline is responding in a timely way to the DCSM content, and (4) ensure that the specialty requirements integration process is working. This requires energetic pursuit on the part of the specialty engineers and eager interaction on the part of the design engineers. All too often both parties tend to withdraw from this interface. What is required is that both plunge across the abyss and engage in real teamwork.

Particular interest must be focused on identifying and resolving issues corresponding to constraints which are identified on the design constraints scoping matrix by a letter "Z".

7.4.2 Specialty constraints communication

Many specialty engineering requirements are contained in often extensive applicable documents. These are imported into the program through a specialty engineering requirements statement such as, "Pressure vessel design shall conform to MIL-STD-1522A." This simple statement places a constraint on the responsible designer to comply with the many requirements in sections 4, 5, and 6 of that document to the extent that they apply to pressure vessels if they are not tailored out.

One of the most difficult tasks in system engineering is the efficient communication of the meaning of the contents of the pile of applicable documents referenced in specialty engineering requirements statements to the designers. Designers who have a great deal of experience, will have acquired a good understanding of many of these requirements, but not all. Recent college graduates will have little experience with the content of any of these documents nor the concept of importing requirements through the referencing mechanism.

Each engineering specialty discipline must work assertively and interactively with the design community to insure that the designers understand the specialty requirements and their consequences. This interaction could take the form of any combination of four principal initiatives on the part of each specialty discipline: (1) specialty checklist, (2) person-to-person discussion, (3) organized interaction meetings, and (4) participation in trade studies and engineering review boards. This very difficult process will eventually be replaced by integrating specialty requirements into computer aided design packages.

7.4.2.1 Checklist approach

In the checklist approach, each specialty discipline which invokes an applicable document containing a voluminous list of design constraints, must translate that listing into a checklist focused on the specific product line and element, if possible. The checklist must be formatted such that the designer can clearly and rapidly understand the important attributes that must be satisfied. This checklist must be made available to each design principal engineer affected by the requirements. The checklists could be communicated with printed material or via a networked computer system.

Each engineering specialty should create a specialty checklist that is responsive to the applicable document(s) called out for that discipline. The checklist simply lists each specialty requirement contained in an applicable document and provides a space for each item to be checked off. A specialty checklist may have several columns for different kinds of elements (AGE, flight vehicle, etc.). The checklists should be created during the early period of the program when the top level system elements are being synthesized.

Specialty requirements often apply to all system elements across several layers of architecture but can best be dealt with in association with assembly or component level design work. The checklists must be available by the time it is necessary to identify requirements for component level elements that will yield to detailed design.

If the checklist approach is applied to all specialty disciplines, there are many disciplines on a program, there are many requirements per discipline, and all of the specialty disciplines use the checklist approach effectively, the design engineers can easily become completely overloaded.

It is, however, important to make every reasonable effort to help the design engineers understand the complete set of requirements that must be complied with. One element of that assistance could be a Joint Checklist Peer Review. This is an organized review of all of the specialty checklists by the specialty disciplines. A representative from each specialty discipline reads all of the checklists of the other disciplines with an eye for spotting any conflicts with their own checklist.

One way to implement this review is to meet for one hour per day for "n" days (where n equals the number of specialty disciplines on the program). Each day, a different one of the specialty disciplines should act as the host, providing copies of his/her checklist, explaining the content, and reacting to critical comment. A system engineer should set up these meetings and oversee them as necessary to ensure issues are clearly resolved. The resultant joint checklist can then be briefed to some or all of the designers with follow-up interaction on a personal basis by the specialty engineers.

Another checklist simplification approach is for each specialty to partition their aggregate checklist into subsets that provide all of the constraints related to particular kinds of elements common to the company product line or particular program elements. This may eliminate many checklist items from the list that a designer interested only, for example, in valves must use. This becomes very important when you consider that a designer may have to deal with checklists from ten or more specialties.

By now you should completely understand how the system development process became so difficult through progressive specialization without concurrent improvements in integration of the specialty engineering views.

7.4.2.2 Individual person-to-person

In the person-to-person approach, each specialty discipline interacts on a one-on-one basis with each designer, with or without benefit of a checklist, to help them understand their requirements. If checklists are used, these conversations can focus on the checklist items and whether or not the designer understands the listed items. Where checklists are not used, the specialty engineer must develop a feeling for whether the designer understands the raw content of the applicable document or allocated specialty requirement.

In these discussions, the specialty engineers may be able to offer advice about which potential design solutions under consideration would have the better result in satisfying their requirements. The specialty engineer may also give the designer some examples from similar programs about how these requirements were satisfied.

7.4.2.3 Organized interaction meetings

The designers and specialty engineering representatives could be brought together periodically for informal discussions about current problems and shared insights. This may be most useful when a development team must respond to a specialty engineering discipline very uncommon to their experience. For example, if a private aircraft company won a U.S. Army contract to provide an air vehicle that could survive in a nuclear battlefield situation, they would have to master some very uncommon design solutions. A periodic meeting may be very effective in this situation until the designers have come to understand appropriate design approaches. These meetings may be more a matter of instruction at first eventually turning into effective technical exchanges.

7.4.3 *Decision support*

The final way the specialty engineers may interact with the design development process is to participate in trade studies, design reviews, and engineering review boards. In these forums, the specialty engineer is responsible to ensure that the preferred solution offered or the decision arrived at in the selection among alternatives has properly taken into consideration the relative valuation offered by the specialty engineer and any specialty concerns.

7.4.4 *Specialty design assessment*

From the 1950s through the 1980s many companies allowed growing specialized engineering design, analysis, and specialty disciplines to form departments and build walls such that development work progressed in a serial fashion. Specifically, specialty groups were allowed to review finished drawings without concurrent interaction during the creative concept and design development process. This phenomena is called stovepipe engineering by some people.

As discussed earlier, the specialty and analysis groups must be employed in a concurrent way in order to create around the creative design engineer the effect of a single, complete all-seeing engineer effective in initially designing-in all product specialty engineering requirements simultaneous with the design of the product manufacturing and employment process. This requires the formation of effective, physically collocated teams of design, specialty, and analysis engineers and their manufacturing, logistics, and operational specialists.

Even in an effective concurrent design environment, however, we should not fail to independently and formally assess the design for requirements compliance. If a project organizes by integrated product development teams (PDT), this can be done through cross-PDT assessment of each others designs to preserve a degree of independence or by a system level specialty engineering function within the PIT. The results from this assessment should be captured in some enduring way so that it may be used as specialty requirements verification at a customer-mandated functional configuration audit (FCA). If checklists are used, a set of signed-off checklists may be adequate for this purpose. This checklist could have one column for pre-design interaction and post-design assessment.

7.4.4.1 *Non-compliance identification*

During the design process, the engineering specialists must study the preliminary sketches and detailed engineering drawings offered for check and release to assess the degree of compliance with their requirements. If a checklist was used as a means to inform the design engineer the specialty engineer should use the same checklist as a systematic aid in the assessment.

The specialty engineer must have some understandable rationale for a conclusion of non-compliance. It is not enough for the specialty engineer to simply conclude that a design fails to meet the requirements. The designer must be offered an understandable statement of the reason in terms of the requirement. A checklist will help in this regard where the specialty engineer may refer to a specific checklist item and describe the design characteristic that is at fault.

It is less costly for the specialty engineers to identify non-compliance issues before design release than after and the earlier the better. But, situations do arise where it is necessary to view several drawings and associated analyses before the specialty engineer can form an educated opinion about compliance. In these cases, some drawings may have been released in the process of waiting for the others needed for the analysis. This is an unavoidable problem the severity of which can be reduced if the specialty engineer gives these kinds of compliance assessments top priority when the complete set of data does becomes available.

7.4.4.2 Non-compliance correction

When the specialty engineer finds a design that fails to comply with a valid specialty requirement, that engineer must first try to resolve the issue with the designer. It will be helpful to refer to the specific checklist item or documented requirement and the specific way in which the design fails to satisfy the requirement in this conversation. The experienced specialty engineer can also offer fairly specific ways the problem could be overcome in the design. The specialty engineer should recognize that the designer may very well feel protective and defensive about the results of his/her creativity. A little tact mixed with assertiveness will generally be repaid in this conversation.

If the direct approach to the designer fails to achieve a satisfactory solution, the specialty engineer must bring the non-compliance issue to the attention of the designer's supervisor, manager, or the Chief Engineer. Where the specialty discipline is on the drawing signoff list, recognition of a problem is guaranteed, but correction is not. Where the engineer is not on the drawing signoff list, he/she must ensure that the non-compliant issue does not pass unnoticed. A program issue system can be useful in such cases. A program issue is a special kind of action item that must be resolved by the designer of the element against which the issue is written. Like all engineering problems, the engineering review board is the court of last resort for these issues.

7.5 Specialty engineering activities

This section offers the system engineer a very brief summary of several specialty disciplines that may be required on a program. Not every one of these disciplines will be required on any one programs. During the program planning process we must understand which specialty disciplines will be required as a function of the nature of the system, the maturity of the related technology, and the degree of correlation with past solutions to this or related problems.

The SOW for a given program should include an identification of the specialty disciplines required. It is possible that a program may conclude that one or more specialty engineering disciplines must be applied even though a Statement of Work does not specifically call for it in order to ensure the design is adequate for the requirements defined in the customer's system specification.

7.5.1 Reliability

Reliability is a measure of the probability that an item or system will continue to function for a specific duration and under prescribed conditions. It is measured in terms of the probability of failure or the mean time between failures in hours. The reliability engineer allocates the system reliability figure to lower tier items in the architecture forming a reliability model. The design team fashions a design that satisfies this allocated figure which is verified by assessing the reliability of the parts and computing the resultant reliability figure for the item as a function of the way the parts are connected and used. Part and component reliability figures are commonly extracted from reference documents listing proven reliability figures for specific kinds of components.

7.5.2 Maintainability

Maintainability is a probabilistic statement of the time it will require to repair a failure. This can be stated in terms of the remove and replace time, total repair time, or other parameters. The maintainability engineer allocates system level repair time to items in the system and tracks design team performance in responding to these allocation. As design alternatives are evaluated, the maintainability engineer looks for features that will encourage or deter maintenance actions.

7.5.3 Availability

Availability is a measure of the probability that the system will be available for use at any point in time. It is measured in terms of one of several particular combinations of system reliability and maintainability figures.

7.5.4 Survivability and vulnerability

This may include nuclear, biological and chemical analyses as a function of the threats posed by a hostile force. The effects of these agents is defined for the benefit of design teams and design alternatives reviewed for compliance with recognized effective solutions to the problems posed by the agents.

7.5.5 Electromagnetic compatibility

On systems that include sensitive electronic circuits that could be upset by strong electromagnetic fields or that are capable of generating such fields that could interfere with other systems, the system will be studied for sources of interference and any that are identified will be corrected to within the levels prescribed by law, program requirements, or sound engineering judgment or recognized standards.

7.5.6 Radio frequency management

Systems which include radio frequency emitters, such as radio or radar transmitters, must have their frequency assignments coordinated with available spectrum assignments by one or more controlling organizations.

7.5.7 Electrostatic discharge

Systems operating in the atmosphere are susceptible to a build up of electrostatic charge that, if allowed to reach a high potential relative the surrounding charge, can have a detrimental effect on sensitive on-board electrical equipment.

7.5.8 Human engineering

The human engineer seeks to ensure that design features reflect human capabilities with respect to recognition of critical conditions and ease of actions that must be taken to operate and maintain system items.

7.5.9 System safety and health hazards

The safety engineer identifies safety requirements based on customer needs and cooperates with design and analysis personnel to understand and identify safety hazards to life, health, and property value. The principal approach is to build a model of operation in cooperation with the maintainability and logistics personnel of the system operations and support process and to examine this process for conditions that can cause hazards to develop. The product is evaluated for ways to prevent these conditions from ever developing or ways to control the risks when they do occur. A hazard list is prepared and ways are found to eliminate or control each hazard.

7.5.10 Environmental impact

The system must operate within a prescribed environmental definition. The system and the environment will interact in certain ways and the goal is to minimize the adverse impact of the system on its environment. This is accomplished by understanding the interface between the system and the environment in terms of all materials and energy that are exchanged across this interface. Each of these interfaces is studied for ways to reduce environmental impact.

7.5.11 System security

The system security discipline seeks to identify risks to the system and identify ways that the system can avoid these risks. The threats to the system must first be listed and then each one of these threats must be dealt with in the design to preclude their occurrence from having an adverse effect on system operation.

7.5.12 Producibility

It is possible to design a product such that it is either very easy to manufacture (or produce) or it is very difficult to produce. Producibility seeks to bring about the former condition by close and cooperative work with the design engineers to encouraging the use of processes already clearly characterized and simple design features.

7.5.13 Supportability and integrated logistics support (ILS)

The logistics specialty engineering activity seeks to identify features that will result in optimum supportability in terms of maintenance (testing, servicing, handling, etc.): spares provisioning; training; and technical publications.

7.5.14 Operability

Operability, in concert with human engineering, seeks to optimize the ease of operation of the system and the effectiveness of the system. A process diagram is used as a basis for evaluating the steps needed to operate the system. Human activities needed to operate the system in each of these steps are studied and operability features offered to the development team.

7.5.15 Testability, integrated diagnostics, and built-in test (BIT)

A system will generally require some means of determining whether it is in an operable condition under certain circumstances. This specialty determines an optimum way to accomplish this end through identification of an integrated view of product and support equipment features and capabilities that together will assure effective testing.

7.5.16 Transportability, mobility, and portability

Some elements of the system may require some degree of physical mobility. This specialty discipline determines the character of the needed mobility and defines ways to satisfy these needs.

7.5.17 Mass properties

The mass properties engineer is responsible for ensuring that the design falls within weight and center of gravity (CG) constraints established for the product. The principal method involves allocation of available weight to system elements and monitoring the design process to see that responsible teams and designers remain true to their allocations. A weights table is established that lists all of the system elements and their weights with subtotals and grand total. Weight margins may be established to protect the project from weight growth problems and provide for management of difficult weight issues as the design matures.

The mass properties engineer must also compute the CG of elements where this is a critical parameter. In maintenance situations this data may be required not only for a whole end item (hoisting and lifting, for example) but for various conditions where the item is incomplete as in assembly and disassembly operations.

7.5.18 Materials and processes

Materials and processes seeks to standardize on a minimum number of materials qualified for the product application and a standard series of manufacturing processes appropriate to manipulate those materials. Some materials and processes may require company specifications, written by materials and processes, to characterize them where adequate definition does not exist elsewhere. Designers must select materials from the standard list provided by materials and processes. The standard materials list is the responsibility of PIT. Any materials and processes specifications will be initially prepared by the PDT that first requires its use and thereafter reviewed, approved, and maintained by PIT.

7.5.19 Parts engineering

The role of parts engineering is to assure the use of parts qualified for the application and to standardize on the fewest possible number of different parts. This is accomplished by development of a standard parts list from which designers may select parts. Any suggestions for additions to the list by designers are reviewed to determine if a suitable part has already been identified or an existing listing can be applied to the new application. Some parts may require a company parts specification written by parts engineering. The parts list is the responsibility of PIT.

7.5.20 Contamination control

Contamination control seeks to limit contamination of the product during manufacture by any foreign material generally defined in terms of particles larger than a certain size. Control is exercised by requirements for processing within areas qualified for a prescribed level of cleanliness and special transportation and handling process instructions.

7.5.21 Guidance analysis

If the system includes an element that must move from one point to another with a degree of position precision, this discipline provides requirements that encourage the needed accuracy and evaluates design features to assure that those requirements will be satisfied.

7.5.22 Structural dynamics and stress analysis

This discipline determines the needed strength of structures under static and dynamics conditions under all system conditions. Computer tools are used to model the structure and support structural design personnel in selection of materials and design concepts.

7.5.23 Aerodynamics

If the system involves movement of an element through the atmosphere at speed, it may require an aerodynamics analysis or wind tunnel test to assure that its shape minimizes drag and offers adequate lift to balance weight under all conditions of flight.

7.5.24 Temperature analysis and thermal control

Heat sources and sinks are identified and the resultant temperature of items in time are determined. Involved in positioning and mounting of items for thermal control and elements involved in altering the environment within which items are located.

7.5.25 Design to cost and life cycle cost

Design to cost (DTC) is a technique to encourage cost-conscious behavior in the development team toward the end that the product development cost targets are met. A system development cost number is first identified and this cost is then allocated down through the hierarchy based on anticipated development difficulty. Margins may be assigned to introduce slack cost figures that can be reallocated as problems develop in design. The design to cost allocations will be managed by PIT but the responsible PDTs will be accountable for satisfying them.

Life cycle cost (LCC) combines DTC with cost figures representing recurring manufacturing, operations, matintenance, and disposal costs. It seeks to define the true total system cost over its entire life cycle. This is commonly used as a principal decision making figure of merit in early program phases.

7.5.26 Value engineering

Value engineering is a structured method for finding ways to improve a product after it has entered a production status. It involves the employment of cross functional teams in the performance of trade studies focused on selection of the most cost effective solution to a production problem. This process may be married to a pre-planned product improvement program for the purpose of determining the precise way that the preplanned improvements will be implemented.

Value engineering, when implemented, will be managed by PIT. Each PDT or PMT will be responsible for value engineering of the products for which they are lead.

7.5.27 Other specialty engineering disciplines

We can continue almost indefinitely with our list of specialty disciplines, but in the interest of preserving our forests, let us try to move on to other pursuits. You should understand by this point that we have specialized with very fine granularity in the engineering community for reasons covered earlier related to our limitations and the continuing expansion of our knowledge base. The challenge for the system engineering community is to provide programs with effective methods within a conducive environment to integrate the work of these many specialists toward program common goals.

chapter eight

Program execution

8.1 Program execution controls

In Chapter 9 we will discuss program integration for the case where the current planning data can no longer be efficiently applied in control of the program. This chapter is focused on program integration when things are going according to plan. We arrive at a condition like this through good planning, faithful execution of the plan, and the good fortune to have develop precisely the conditions that we had planned for within an environment of imperfect knowledge of the future. Throughout the period when things are going according to the plan, we need to have our sensors turned up to detect indicators of potential problems or risks, for risk changes over time as a function of what we do, and what happens in the world around us. Chapter 9 will pick up on that point. During this discussion, we will assume that these sensors are reporting nothing but a Go condition.

Our program plan, integrated master plan, system engineering management plan, or top level program planning document by whatever name and program level schedule are the two principal program execution controlling documents at the top level. Others high on the list are the statement of work (SOW), work breakdown structure (WBS), and the contract data requirements list (CDRL). The combination of these documents tells what has to be done, when it must be done, and by whom to what end. Further, they define data products that must be delivered to the customer and provide a structure within which program cost may be managed, collected, and reported to the customer.

8.2 Alas, good planning is not everything

These documents do tell what must happen but they do not, of course, tell what is happening during execution. So, in addition to good planning data, program management personnel need good program status tracking data. Good management skills and judgment are also needed to take advantage of accurate status information to steer the program along a sound course. Program status data must be compared with planning data and appropriate direction determined to encourage continuation on the plan. Most companies have cost/schedule control systems (C/SCS) that relate to the WBS by whatever name. These systems are designed to provide management with financial and schedule data that tells how the program is doing relative to the plan. Many of them at the time this book was published, provided exactly that, data, in reams of computer print-outs.

There is a lot of work for computer programmers to do in refining the presentation of C/SCS information to managers and workers in an easy-to-understand way that will lead to effective decision-making. There are some good examples of a sound direction for this work in the form of the technical performance measurement (TPM) approach and functional management metrics. Most C/SCSC systems are still operating in the batch processing era while the world has moved to on-line access.

Program management must also have access to the third leg of program management information dealing with product technical performance progress. One very effective tool for managing product performance is the TPM process. In the TPM process, customer and company management agree on a relatively small number of quantifiable product performance parameters that will be controlled under the TPM program. These should be key requirements that collectively span the development effort, ones that will be adversely effected when trouble is encountered. The parameter values are tracked in time on TPM charts that are annotated to note critical events related to changes in values. When a reported parameter value exceeds an allowable deviation from planned value, it is clearly visible on a graphical chart. Specific actions are taken to drive the value back within limits. The value continues to be monitored to see if the corrective action has the desired effect.

Weight, maximum speed, guidance accuracy, predicted turn-around time, throughput, and memory margin are examples of possible TPM parameters from several different fields. In picking these parameters it helps to know what kind of product requirements your company has had trouble in satisfying in your history. Some company people may be concerned that if the customer has access to this data it might make the company look bad to use these parameters for formal performance measurement. Some company program managers would prefer to have a contract set of parameters that are not so effective at spotting company problems and additional parameters that are evaluated only internally that may recognize traditional problems. The customer that knows its supplier will not stand for this, of course.

The TPM technique is a closed loop process and that is what program managers need for cost and schedule control as well. In these systems, an unacceptable deviation is visually obvious from the management information, specific action is taken to correct the indicator of unsatisfactory program performance, and the indicator then provides, with some time constant, feedback of the effect of the action taken. TPM and the application of this same approach to C/SCS are examples of what are called metrics. They all work this same way. Key numerical parameters are charted in time and used as a basis for management feedback to the process. The process reacts and the results are displayed by the charted parameter. Chapter 9 offers more detail on TPM.

8.3 Implementing the IMP/IMS

In Chapter 6 we encouraged the use of the U.S. Air Force program planning initiative called the integrated management system using an integrated master plan (IMP) and integrated master schedule (IMS) as the principal management tools. In so doing, we need the information inputs provided by the systems discussed above focused on the events identified in the IMP. The IMP provides not only an organized set of events to which we have hooked the tasks defined in the statement of work (SOW), but for each event a list of accomplishments we should expect to see materialize and specific criteria by which we may objectively judge whether or not those accomplishments can be claimed. The IMS fixes these events and accomplishments in time.

The IMP/IMS should have been organized such that it is perfectly clear which product team is responsible for the work covered. There should be an overall IMP sheet covering each team, that may have to be expanded in detailed team planning data, but from which program management personnel may determine what should be happening on the isolated team and relative to work on other teams.

8.4 Controlling the advancing wave

One of the most confusing and distracting things about the development of a complex system is that at any one time all of the pieces of the development work are not at the same stage of maturity. They cannot possibly be if we are serious about the notion of top-down development since a sequence is implied. As we develop the architecture for our system it expands into the lower reaches as noted in Figure 8-1 originating with the simple allocation of the customer need to the block called system. Subsequent allocations of identified functionality result in identification of lower tier elements and requirements corresponding to those items. During the decomposition process

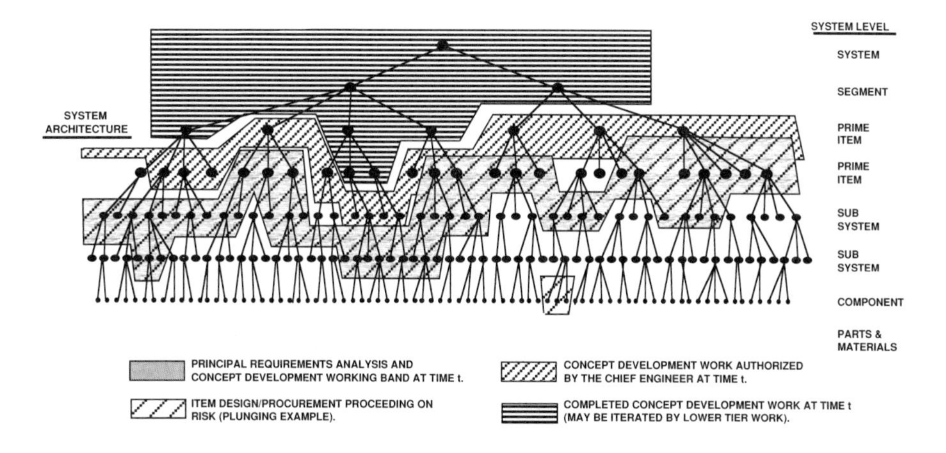

Figure 8-1 The advancing wave.

there is an advancing wave of development downward through the architecture to identify items and requirements for those items.

When this process hits bottom, the wave rebounds upward and we move to the design and integration process. Hitting bottom is characterized by: (1) requirements are identified for everything in the architecture and (2) everything on the lower fringe will surrender to detailed design by an identified specialized design team or associate contractor or it will be purchased at that level. These two strokes or wave movements must be imbedded in our IMP/IMS, but we also need an effective means to track how well we are doing during execution. That means should also help the principal decision-maker(s) to encourage a structured approach defined by requirements approval before design.

Figure 8-2 illustrates one version of a development control matrix designed for that purpose. The matrix tracks the development of each selected architecture item in three primary development modes: (1) requirements documentation, (2) concept development (preliminary design), and (3) design development (detailed design). In each case due dates are identified from the IMS or lower tier supporting schedules. The dates should be included but were dropped here for lack of space. Each development mode portion of the matrix also calls for an engineering review board (ERB) number entry where the item was approved to progress to the next mode and a status column where you could use numbers, letters, or colors.

There are many other things that a particular development program might choose to place on such a matrix. But matrix design is inevitably a compromise. You need to select things that help you to focus on status visually. You want to be able to tell quickly what the status is on every level and branch of the architecture shown in Figure 8-1 down to the lowest level at which you choose to manage. Naturally, the PIT might have a system level matrix and each PDT leader could have a matrix for their item, thus reducing the vertical dimension of any particular matrix.

Another technique would be to eliminate the status columns on the matrix (freeing up space for other text data) and place the status information

ARCHITECTURE IDENTIFICATION		PRINCIPAL			REQUIREMENTS DOCUMENTATION						CONCEPT DEVELOPMENT			REMARKS
ARCH ID	ITEM NAME	CONTRACTOR	TEAM	ENGINEER	TYPE	NUMBER	REV	DUE DATE	ERB NBR	STAT	DUE DATE	ERB NBR	STAT	
A14	COSGROVE UPPER STAGE	GSTS	PIT	JONES	CI	72-1934	-	02-25-90	051	G	03-10-90	063	G	
A141	EQUIPMENT MODULE	GSTS	PDT 2	BURNS	DRD	72-2430	-	03-10-90	062	G	04-20-90	068	G	
A1411	EM STRUCTURE	GSTS	PDT 2	ADAMS	DRD	72-1122	A	04-05-90	066	G	05-10-90	080	O	
A1412	AVIONICS SUBSYSTEM	GSTS	PDT 2	PERKINS	DRD	72-3323	-	04-10-90	067	O	05-15-90	081	R	PLATFORM MOUNTING SCHEME
A14121	GUIDANCE & CONTROL SUBSYSTEM	GFG	PDT 2	SMITH	DRD	72-2579	B	05-20-90	083	R	06-25-90		O	
A141211	ON-BOARD COMPUTER	TELEBOND	PDT 2	BROWN	PS	72-2582	-	06-10-90	091	O	07-10-90		O	CHIP GIDEP ALERT
A141212	INERTIAL NAVIGATION SET	REYNOLDS	PDT 2	WINDHAM	PS	72-2585	A	06-15-90		O	08-10-90		O	
A141213	STAR TRACKER	GFE	PDT 2	FLETCHER	-	S-1267-3	A	02-10-90		G	02-10-90		O	
A14122	ELECTRICAL SUBSYSTEM	GSTS	PDT 2	BLACKMER	DRD	72-9333	-	05-10-90	082	G	06-10-90	093	G	
A141221	BATTERY	RPS	PDT 2	GOMEZ	PS	72-9354	C	06-10-90	090	R	06-15-90	095	O	STRIKE SCHEDULED
A141222	WIRE HARNESS	GSTS	PDT 2	CHIN	NA	72-9365	-				06-20-90			
A141223	POWER CONTROL UNIT	ESI	PDT 2	JOHNSON	PS	72-9202	B	06-10-90	089	G	07-25-90		O	
	STURE					72-1211	-	05-20-90	084	G	06-10-90			

Figure 8-2 Development control table.

on a graphic of the architecture block diagram by coloring the blocks in accordance with an agreed upon code. This would require three different matrices for the three levels of status tracking shown in Figure 8-2 or some scheme to multicolor a block. Either of these techniques can be accomplished on the computer with network access to the graphics.

Whichever status coding scheme we select, we should always see the requirements for an item go green before detailed design work is allowed to move to an in-work status. The author would break up the requirements status tracking reporting into two steps: (1) a requirements list that captures the functional requirements for the item (what does it have to do and how well does it have to do it) and (2) the item specification. Concept development should be allowed to proceed based on an approved requirements list. Design development should not be allowed to proceed without an approved specification.

8.5 Summing up

In summary, that worn out maxim of good management comes to mind, "Plan the work and work the plan." To which the author would add, ". . . while keeping one eye on risk." Risk, causes of program discontinuity that can result in a break in execution in accordance with the plan, and how to get back on track after having fallen off the plan, is what the next chapter will deal with.

chapter nine

Discontinuity management

Perfection is seldom attained. If a program was perfectly planned and executed with good fidelity to a good plan and everything unfolded as conceived by the planners, everything would go according to plan. Has there ever been such a program experience? Probably not. We have to plan programs with imperfect knowledge of the future and this reality practically guarantees that everything will not go precisely according to plan. Your competitors will also remain actively tuned to your progress and work to counter your best efforts with new products and advertising. Therefore, we must have a way to detect that we have fallen, or are in danger of falling, off the plan and the machinery for getting back on. We must have a capability for quality re-planning of remaining work. Perfection can be approached through good planning as discussed in Chapter 5, and good management leading to good execution but it is just very hard to know everything that will happen during program execution at the time you are making your plan. We must plan a program with imperfect knowledge of the future.

9.1 Discontinuity defined

A program discontinuity is an interruption in planned activities. It is a condition that precludes continued efficient execution of work in accordance with the predetermined plan. There are three principal factors involved in determining whether conditions of discontinuity will develop during program execution:

a. Planning quality
b. The degree of correlation between the assumed conditions that will be in place during execution while planning and the reality during execution
c. The fidelity of plan execution

Figure 9-1 is a three dimensional Venn diagram illustrating the possibilities from these three factors. In the simple case of binary possibilities for each variable, there are two cubed, or 8, outcomes. The preferred outcome during execution is, of course, that we had planned well, executed the good plan faithfully, and conditions in effect during execution were those for which we had planned. All of the other combinations will yield less desirable results.

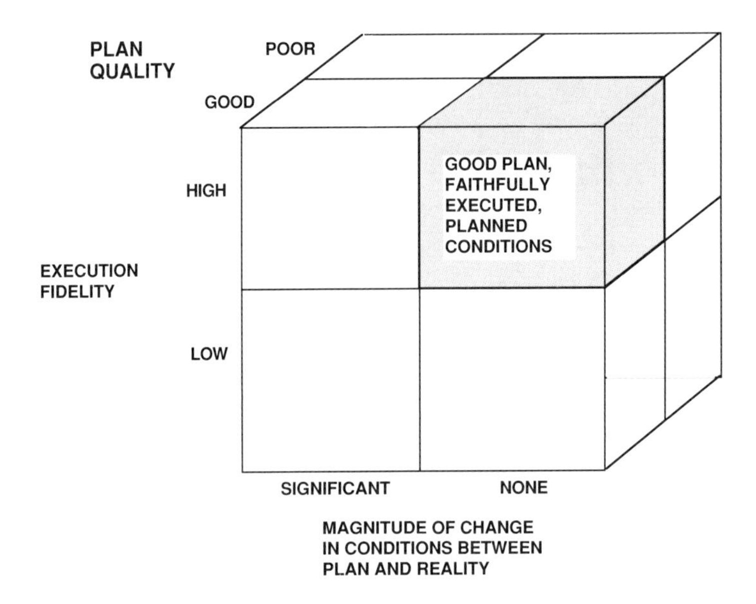

Figure 9-1 Discontinuity cause cube.

One possible exception is that a failure to execute a bad plan faithfully amidst radically different circumstances than planned could conceivably turn out very well. This is a condition of almost total discontinuity, however, involving instinctive management without a plan that requires a lot of good luck and unusually fine intuitive management skills. The reader will have to look elsewhere for guidance for this possibility. If the management team is this good, think what they could do with a good plan.

Given that we plan well and execute the plan well, the only two sets with which we should have to deal in Figure 9-1 are the ideal situation and that corresponding to conditions that have changed from those planned. This may be more than we should expect out of life, but you will have to judge for yourself based on your own organization's history.

We generally think of discontinuities as being bad, but it is possible that conditions can change for the better and we may wish to change our plan to gain full benefit from these changed conditions. For example, our plan could involve using a very expensive technology because another is thought to be too far out in development. During our program execution, that other technology might become available in practical terms making it possible for our customer to save hundreds of thousands of dollars. Certainly, we would want to embrace this new technology no matter what our current plan calls for.

9.2 Discontinuity detection

A program discontinuity can arise from several sources suggested from Figure 9-1 and we need a detector tuned to each of these sources. The three principal program planning focuses are technical or product performance, cost, and schedule. Most everything that can go wrong in terms of plan quality, execution faithfulness and changed conditions can be expressed in these terms. Some people, including the author, would add technology to this list even though technology unavailability can be expressed in terms of cost, schedule, and performance. A new development program that is pushing the technology will benefit from this added trigger. A program that can rely on existing technology need not include it. This gives us four triggers for which we must arrange detectors. For particular programs, or just because of historical precedent, a particular company may choose to add other specific triggers to this list.

The detectors we need are really tuned to what commonly are called risks. Risks are potential program discontinuity triggers that should be avoided, mitigated, or corrected for. Where ever possible, we should avoid them by detecting their potential appearance in the future and taking action in the present to ensure that the conditions needed to bring them into existence never occur. When a risk is upon us, but not yet disrupting our plans, we need to be actively mitigating it to prevent it from becoming an open discontinuity. A risk that is fulfilled is a program discontinuity that must be removed through re-planning.

9.2.1 Cost and schedule triggers

The Department of Defense (DoD) requires its contractors to use a certified Cost/Schedule Control System (C/SCS) for tracking and reporting program cost and schedule information. The information these systems produce is adequate to detect possible or realized discontinuities from these two triggers. As noted in Chapter 5, however, most of these systems are still functioning in the batch processing mode and need to be updated to permit on-line access and lower tier organizational reporting.

Figure 9-2 shows one function of the PIT to be Program Cost/Schedule Control. This responsibility could, of course, be assigned to the program business team alternatively. This activity must constantly monitor the performance of all of the PDTs and the PIT against planned expenditure and achievements using the C/SCS. Deviations beyond predetermined boundary conditions trigger a potential risk that has to be studied by someone responsible for program risk analysis.

9.2.2 Product performance trigger

DoD customers also encourage the use of a technique called technical performance measurement (TPM) to detect problems in satisfying the key technical product requirements. TPM involves customer and contractor agreement on a small list of key quantified parameters, parameters the condition of which signal the general health or illness of the complete program and product system. Weight is a common parameter chosen where this is an important

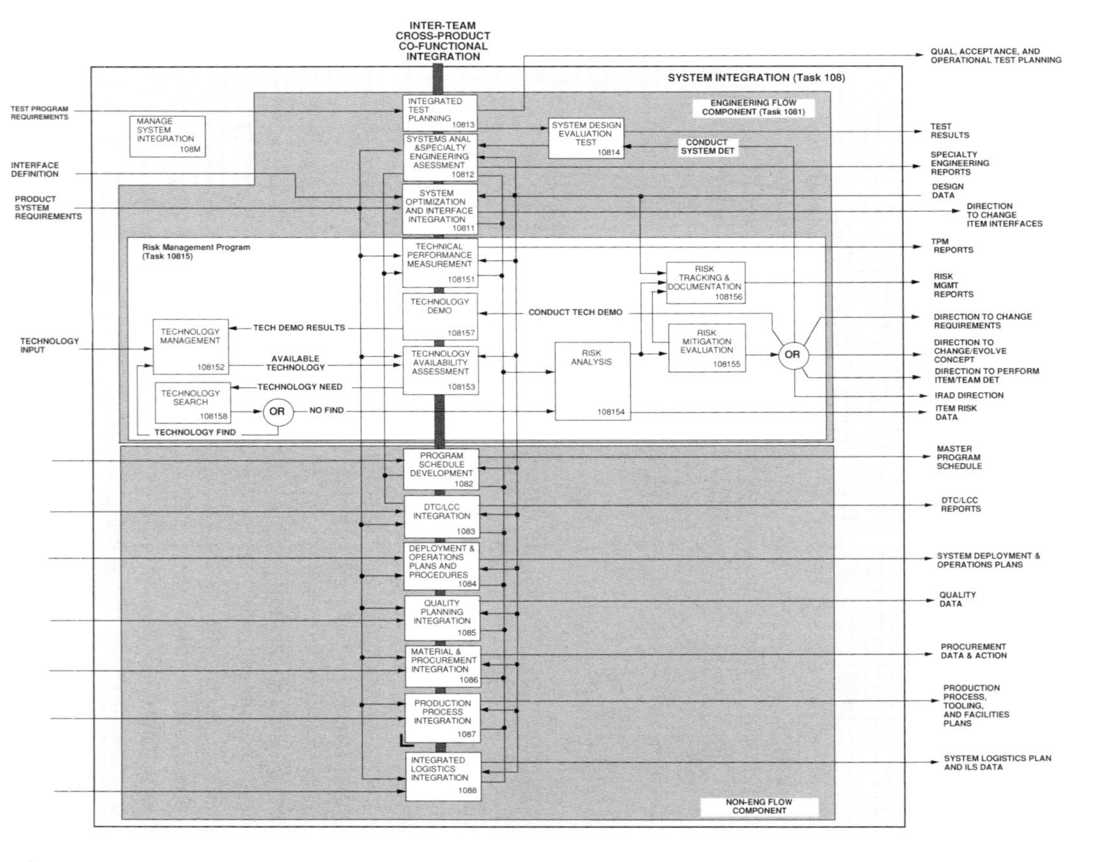

Figure 9-2 System integration process.

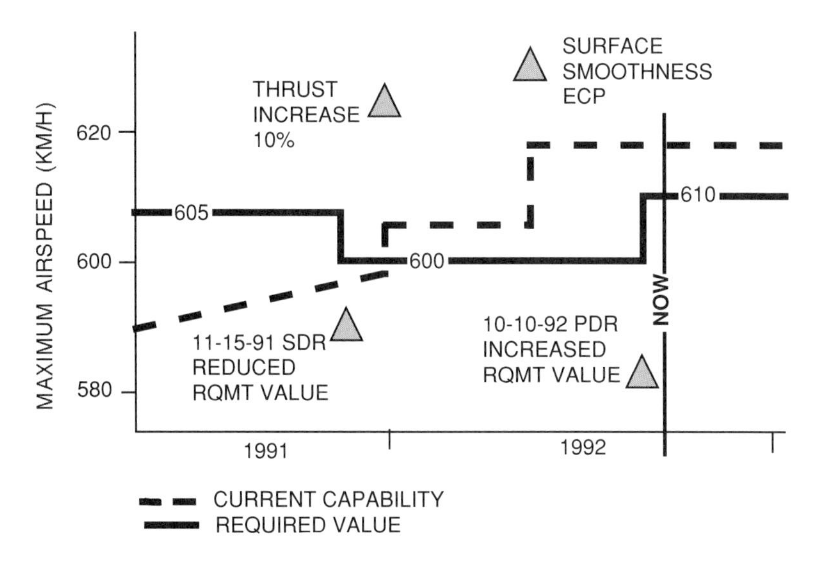

Figure 9-3 Typical TPM parameter chart.

characteristic, such as in aircraft systems. The contractor will track the value of these parameters in time against the required value and make predictions, based on planned work, how the value will change in the future relative to the required value.

The history of past values and future predictions of these parameters is reported periodically to the customer in graphical form. If a TPM parameter is outside predetermined boundary conditions, the customer will expect the contractor to explain the reasons and what they are going to do about getting the parameter back on track with its planned value. As noted in Figure 9-2, the PIT TPM function feeds its concerns to the person responsible for risk analysis.

Figure 9-3 illustrates a typical TPM chart showing the requirement and current value historical traces. When the current value or predictions of future value exceed the agreed boundary conditions, the contractor will be required to brief their planned actions to bring the parameter on track. Boundary conditions are defined to permit the contractor some slack within which they may manage the program without undue customer micro management.

9.2.3 Technology trigger

There is no comparable highly organized mechanism in common use for technology triggered discontinuities. It is not difficult, however, to invent one here and now. First we need to know what technologies our design will have to appeal to. Each development team or principal engineer should be called upon to survey needed technology for the planned design concept. Concurrently, the manufacturing, quality, logistics, and operations people should be doing the same thing for their own fields based on the current product and process design. We then must compare the content of these lists against a list of technologies available to us. These technologies may be available to us because: they are freely available for anyone's use, we know how to gain access to them on the open market, or we hold them as proprietary properties.

There is another class of technologies that drive program managers to insanity that are said to be just over the horizon. If we are depending on one or more of these coming in during our program, we are probably setting ourselves up for a grand discontinuity by plan. All immature technologies should be recognized in our risk management concerns and accompanied by risk mitigation plans corresponding to the case where they do not come on line at the needed time.

As the first gear in our mechanism for a technology trigger, we should require design groups (manufacturing and logistics as well as product) to identify the technologies they intend to appeal to for each design concept during preliminary design. The system engineering community (in the form of the PIT) should review this mix against available technologies. as indicated in the Technology Availability Assessment block of Figure 9-2. The program must, first have a way of knowing what technologies are available that might have a program relationship and this may require a specific technology search or simply the application of an edited list from a comparable prior program.

In all cases where there is a match between required and available technologies, there is no triggering condition present. Where it can be demonstrated that a technology needed by a PDT does not now exist on the program list or there is a likely probability that it will not be available in a practical sense at the time it will be needed on the program, there exists a technology discontinuity trigger event and corresponding risk. The appropriate response is to conduct a technology search followed either by adding a found technology to the available list or signaling a technology risk to the risk analyst.

The outcome of the risk analysis process may be direction to the designer to change the design to avoid the unavailable technology or a technology demonstration to develop and report the new technology needed for the existing design concept. The later path should carry with it a periodic monitoring of the status of this new technology as the time approaches when it must be mature. If there is a risk that it may not be available in a timely way, it may be necessary to run parallel development paths to mitigate the risk of the technology not coming in on time. In this case, it may be possible to convince the customer that the alternate technology will be adequate for initial capability, even though it does not meet the system requirements, and that it can be replaced at a particular cut-in point in production. The customer may agree to a deviation from specification requirements for a particular number of articles that will be corrected when the needed technology becomes available.

9.3 Risk assessment and abatement

When any of these triggers are activated for a specific risk, they should stimulate the PIT and/or program management to a particular course of action. First the risk should be studied to determine the nature and scope of exposure. What specifically will happen if the risk becomes a reality? Then we must try to ascertain the probability of the risk materializing . A risk with a combination of very serious consequences and a high probability of occurrence should be dealt with as a serious matter. A risk with minor consequences and low probability of occurrence may be set aside. In between these extremes we need a policy for selecting risks for mitigation. The problem is that we have limited resources and cannot afford to squander them on every possible problem.

It is helpful to have a special form or worksheet for evaluating risks that encourages the person most knowledgeable about the risk to provide the information needed by those who will decide whether to spend scarce resources on mitigating it. B.S. Blanchard, *System Engineering Management*, Wiley, 1991 and Department of the Navy publication ONAS P 4855-X, *Engineering Risk Assessment*, August 1985, offer examples of worksheets for this purpose.

The program should also have a means to communicate in a summary way what risks are being managed and what their status is. Figure 9-4 illustrates a fragment of one graphical way of doing this developed or adapted by Ms. Rikha Patel, a system engineer at General Dynamics Space Systems Division for the Advanced Launch System (ALS) program. Each major program area is identified in a hierarchical structure (this could be the WBS) with subordinate areas in each case listed. For each block on the diagram, three blocks corresponding to the three major risk types (cost, schedule, and technical), offer the analyst a place to enter the corresponding risk probability. The latter are given in colors as: GREEN=Low Risk (L), YELLOW=Medium Risk (M), and RED=High Risk (H). Letters are used here to avoid the need for colored pages and can be used in databases.

The PIT and program management use the status summary to determine on a program scale where the most serious risks are and how to allocate available budget in the most intelligent way for the best overall effect. Each active risk must have a principal engineer assigned to work issues associated with that risk and take appropriate mitigation actions. Periodically the PIT, program management, or a special risk management board should meet to assess the current status on all active risks and provide direction for future work on those risks. Where a risk has been fully mitigated, energy should be focused in other directions. As new risks become identified and determined to be serious, they should be formally accepted into the set being actively managed.

9.4 Formal discontinuity identification

At any one time, a program may be managing 25 risks using the assessment summary illustrated above. If the risks are well mitigated and conditions

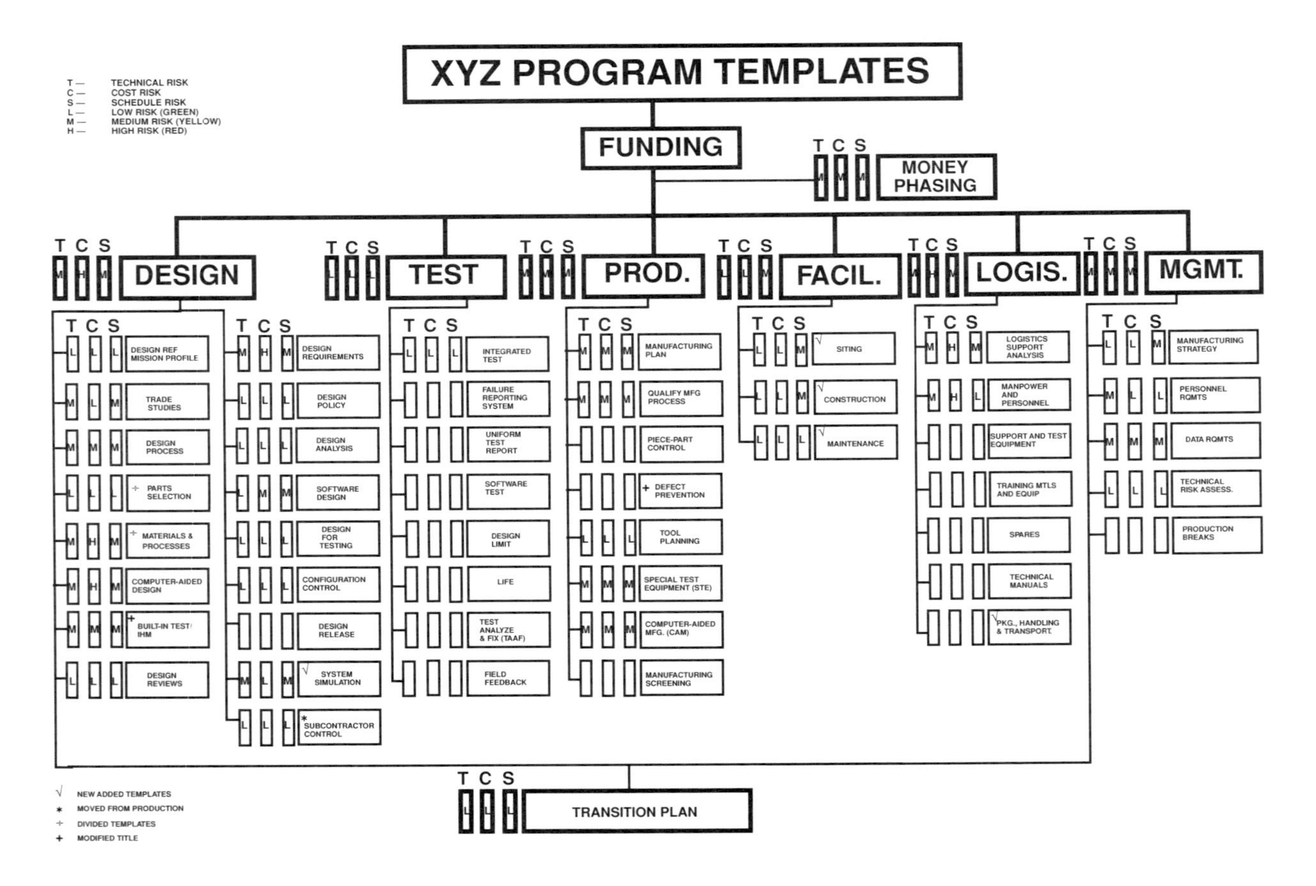

Figure 9-4 Program risk assessment status summary.

permit, none of them may ever become program discontinuities requiring program re-planning. At the same time, if we have done a good job of monitoring the discontinuity triggers and converting them into risks to be mitigated before reaching a discontinuity status, no discontinuities should befall us that do not flow from our on-going risk management program. An exception would be an act of God like an earthquake tearing our factory in half or a flood carrying away downstream the factory of a prime supplier.

These are cases where a discontinuity materializes instantly without passing through our best efforts to prevent them from happening. If we include every possibility, however remote, in our risk mitigation work, there may be insufficient resources to do the actual program work, so some unforseeable risks will remain despite our best efforts.

Our periodic review of the risks that are being managed should ask if any of these risks have reached a condition where we can no longer manage related progress in accordance with our current plan. This could be another indicator that is tracked on our chart. When we must answer this question with a yes, we must accept the need to re-plan part or all of the program depending on the scope and seriousness of the discontinuity.

9.5 Program re-planning

We can apply the same techniques in re-planning that we applied in the original planning activity. Clearly, we must identify the scope of the tasks impacted by the discontinuity. These tasks, or an alternate set of tasks, must be defined in the IMP format as they relate to major program events. The interfaces between these tasks and other program tasks must be studied to determine if the critical path is influenced possibly triggering other schedule risks not previously identified. If possible, the original major events schedule should be respected unless the customer indicates otherwise. This may require parallel rather than serial work performance or overtime to avoid overall program schedule impact. Unless the schedule was very optimistic in the first place or prior re-planning has absorbed all of the slack, there may be some margin in the schedule that precludes major program rescheduling.

chapter ten

Generic program preparation & continuous process improvement

10.1 Being prepared is better than not

The history of the World and mankind leads to the encouraging conclusion that we humans strive for the better in the long run. Nature applies the powerful engine of environmental adaptation and survival of the best prepared in a chaotic fashion leading to improvements in the survivors. This process takes a relatively long time and those in business do not have the time or money to wait around for the death of ill-prepared competitors. We have to take conscious, positive action continually to ensure we will always be in the set of survivors. Humans have the advantage of understanding the notion of survival of the fittest and can influence the outcome by constantly improving their capability and readiness. The phrase continuous process improvement is an expression of this same notion.

In a business environment in a free market economy we survive by the quality and value of our products, relative to our competitors, as perceived by our customers. It is therefore of interest to us what the customer thinks of our product. This fact may be lost in the discussion in this Chapter so let us begin by ensuring that it is a part of our understanding.

A company needs a feedback mechanism giving it valid information about customer attitudes. This may be the sales force, a field engineering group, a marketing team, a customer relations office, or the program manager. Regardless how it is implemented, this communication channel must exist and have influence on changing the way we do business.

The principle topic in this chapter is the exposition of a method whereby a company may continuously be preparing itself for better work in the future by monitoring its present performance, learning from that performance, and making small adjustments in methods that lead to improved product quality and reduced costs that can be passed on to customers as better value that will encourage customers to return for more.

Where the customer is a large institution like DoD, NASA, or DoE, the voice of the customer includes understanding their model of system development and product requirements standards described in thousands of documents called applicable documents. This name is derived from customer reference to them in specifications and statements of work. They are referenced in these contract documents as applicable to the product or process and their content must be complied with under the contract just as the direct content of the system specification and SOW must be complied with. We will see in this chapter that this mass of documentation, for those who must abide it, contains within it a diamond of great value to those who wish to be counted among the survivors.

The coverage in this chapter of customer applicable documents, and their connection to continuous process improvement, may be lost on those only guided by their own commercial instincts, but the same basic notions apply. Instead of using the applicable documents identified by our customers for guidance in improving our processes we may have to rely on more direct feedback from our unorganized customers or reference to commercial standards.

10.2 Continuous process improvement using metrics

First, let us trace the outlines of a general plan for maintaining readiness to respond to our customer's needs and providing our company with an environment for continuously improving our performance while reducing cost. Figure 10-1 offers a blueprint process applicable to many companies. Commercial enterprises may not have to deal with the proposal process but have to deal with some business acquisition process that can be substituted in place of it.

There are two very important signals we need to be tuned to when trying to improve our process. The first is the voice of the customer, as discussed above. The second is our own internal voice. Very valuable information is available within our own organization and may be going down the drain. On-going work on current commitments provides sure knowledge of a company's product quality and cost picture. These signals will be lost, however, if we have no way to capture them. It is necessary to measure our current performance in some intelligent way to acquire the data needed to characterize current performance and the effects of future changes.

Measurement requires a set of metrics that will produce useful numerical figures of merit about our process. These metrics should be automatically produced in the process of doing work. Avoid installing a parasitic metrics manager and staff. This should be a standard practice of good management and managers. We said in Chapter 3 that the functional axis of the matrix should be responsible for the company methods, tools, and availability of qualified personnel. Therefore, functional management should be charged with implementing appropriate metrics for monitoring program performance of their processes. These metrics provide functional management with lessons learned from current work that must be analyzed to determine what if any action should be taken to improve performance. Incidentally it will also provide the general manager/CEO with information about the relative performance of the several programs in-house.

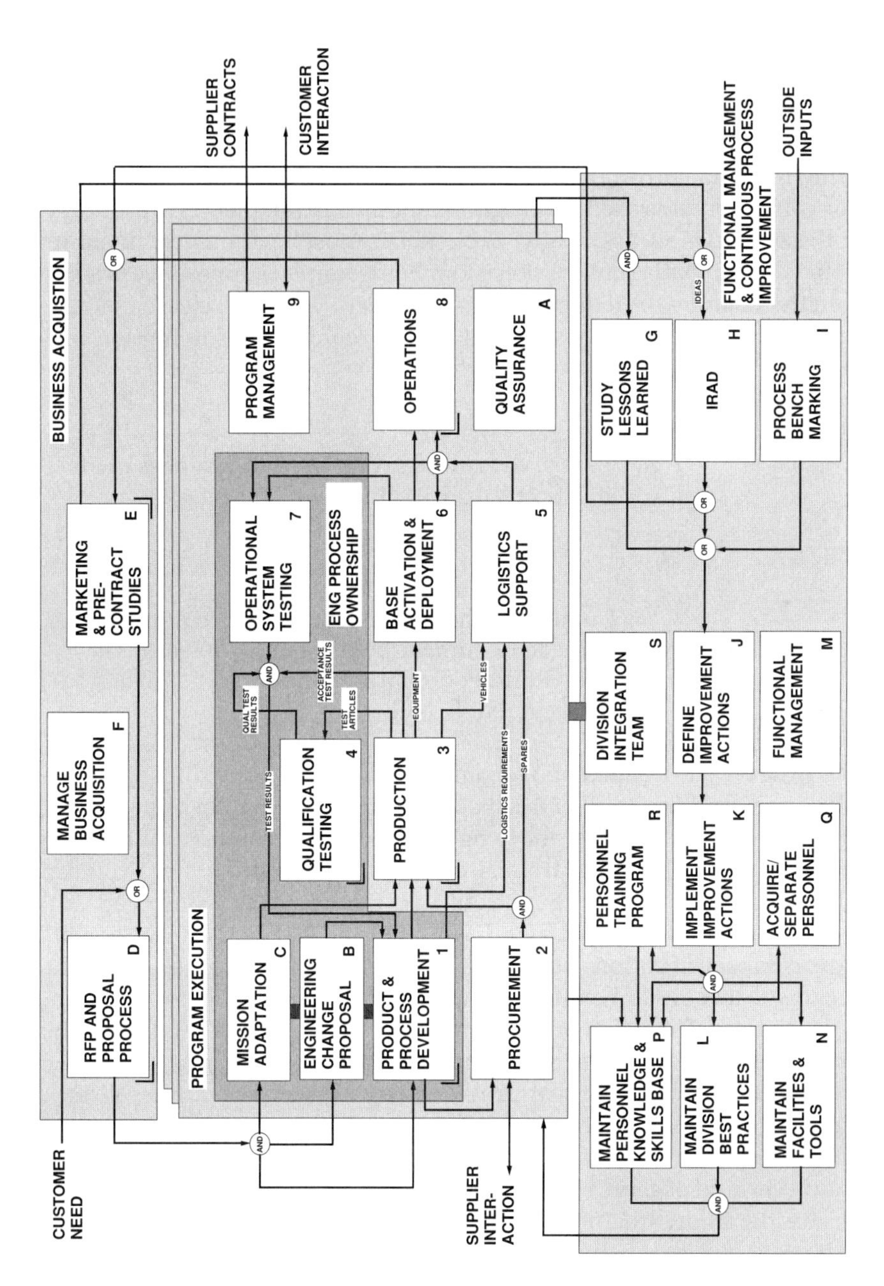

Figure 10-1 Continuous process improvement environment.

Where metrics indicate poor performance, it is suggestive of either a need for improvements in methods, tools, or personnel qualifications or improvements in program execution and management. Functional management should be called upon by programs to participate in design and process reviews as a means of gaining insight into program execution such that all of the information needed to decode lessons learned data is accessible to functional management.

Each functional department should have a metrics list and a means for capturing the data corresponding to these metrics from each program. Ideally, the computer tools used on programs should be designed to capture and report these metrics automatically. Each metric must be numerical in nature and defined with mathematical precision. For example, suppose we wish to measure the volatility of requirements subsequent to initial release of specifications where low volatility suggests a good requirements definition process. Lets define a metric to measure this.

METRIC NAME: Requirements Volatility (V)

DESCRIPTION A measure of the cumulative number of changes made to a specification subsequent to its initial release.

DEFINITION $V_i = \frac{C_i}{R_i}$, where

C_i = Cumulative number of changes to requirements made to document D_i since initial release.

R_i = Total number of requirements in initial release of document D_i.

The figure V_i, as defined above, only applies to an individual specification. We can extend this to all of the specifications on a program by averaging similar figures from all specifications on the program. A company or division figure can be derived by similarly combining the program figures. You can probably imagine that this figure could be automatically computed and tracked by a requirements database. Each time you change the requirements for a particular specification subsequent to an initial release date stored in the system, this metric could be incremented and always available for review at any point in time. The value of the metric over time could be plotted by the system. The author was aware of no computer tools on the market at the time this was written (early 1993) that provided these metric support features.

The figure may not have any significance initially because you have no experience upon which to base a conclusion. If you track this metric over time, however, and observe changes that result from specific actions you take to improve the figure, the metric will start to provide value. These metrics can be tracked in time, as shown in Figure 10-2, exactly like technical performance measurement (TPM) parameters used to monitor program technical performance. Like TPM parameters, you want to select a limited number of functional metrics each of which is useful in detecting quality of performance

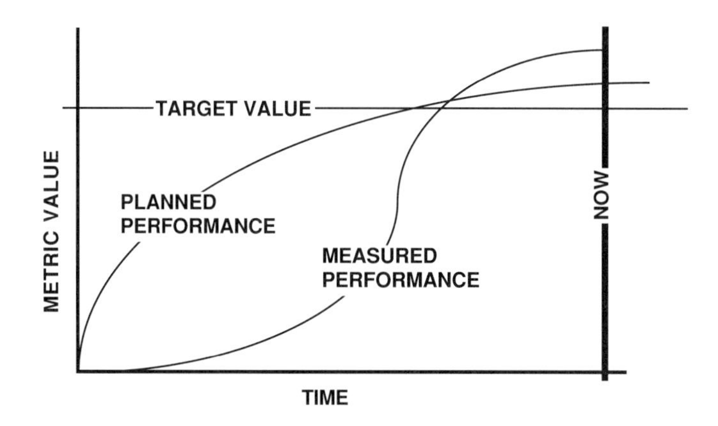

Figure 10-2 Typical metric tracking chart.

in specific activities and together give you a comprehensive view of performance. It is almost as bad to have too many metrics as to have none. To manage is to focus on the important. Too many metrics makes it hard to focus on the important.

Metrics development should be accepted as a long term process. Eventually, you should have a generic set of metrics monitoring performance on all existing programs. Any new program will be formed with clear knowledge of what is expected in the way of functional monitoring of their performance. It is sad that in many companies, management is only concerned about how the customer measures their performance at periodic reviews and they show no interest in any internal measurement process. Your company has a right to a unique identity apart from your customer base and that is the fundamental premise of the remainder of this chapter.

10.3 Generic preparation

Figure 10-3 illustrates an overall process for the RFP/Proposal Cycle. We may enter this activity following identification of a new need by the customer or as a result of encouragement of the customer by a contractor based on new insights into the customer's problems. We may also enter through interphase cycling on a contract won for a current or prior phase. In the latter case, we may already have performed one or more prior phases and have a considerable database upon which to build. The intensity with which we pursue some tasks identified on Figures 10-3 through 10-5 will depend on this past history as well as the customer's request for proposal (RFP).

The scenario expressed in this section may not relate closely to the problems faced by a commercial enterprise which sells directly to the public in that there may be no RFP or proposal involved. Many of the other components can prove useful, however, as a means to continuously prepare for and implement programs producing greater value for customers. For those interested in DoD or NASA customers, you should take special note of the

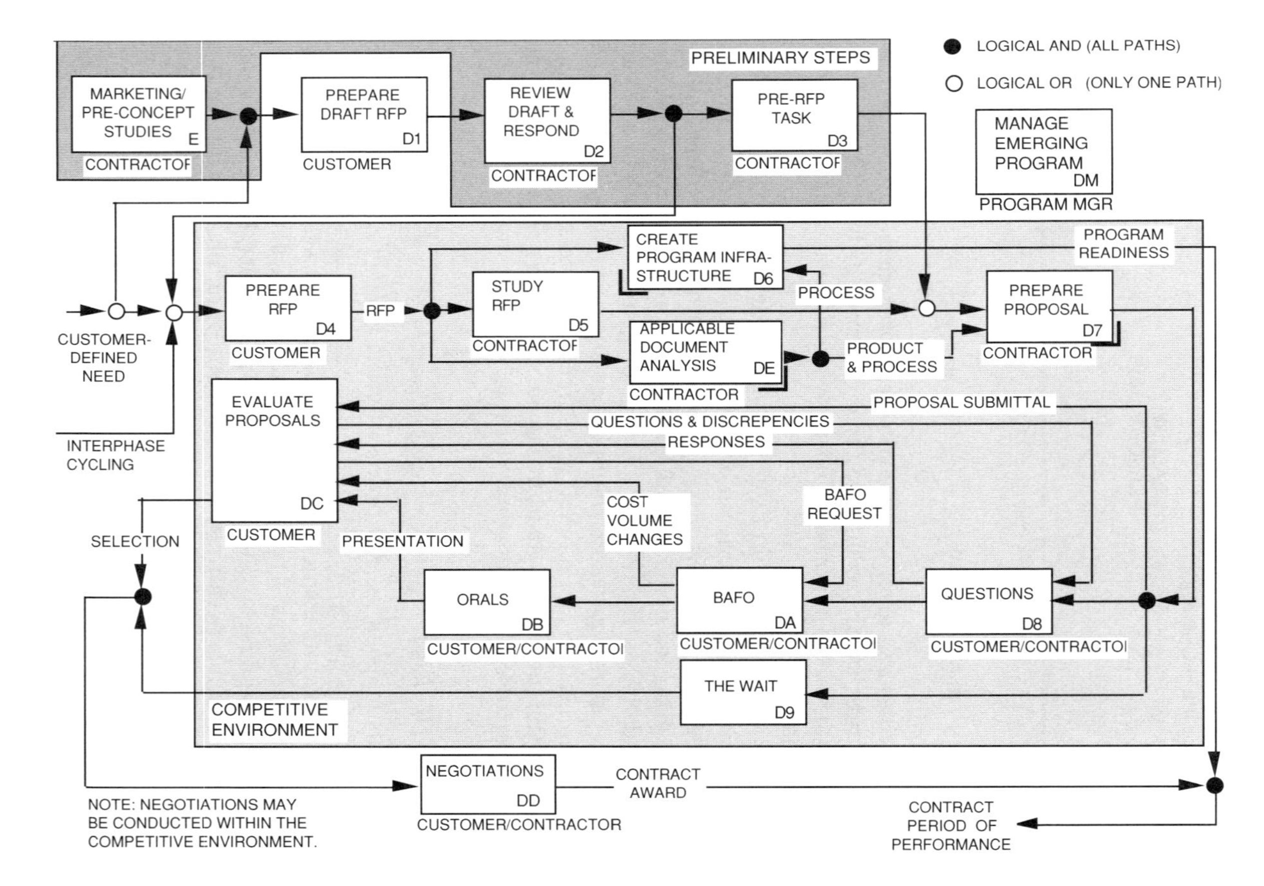

Figure 10-3 RFP-Proposal cycle process diagram.

encouragement to create the program infrastructure concurrently with preparation of the proposal such that you are ready to execute when you win. The author is reminded of the euphoric feeling felt by all on the General Dynamics Convair Advanced Cruise Missile team when the customer announced GD had won the competition. It went something like this, "Wow! We won! .... Oh my God, we won."

The preliminary steps in our scenario involve customer preparation of the RFP, which may include issuance of draft copies to prospective contractors for review and comment. Following receipt of the formal RFP, the contractor must study it and reach a decision whether or not to bid. Given an intent to bid, the contractor must strike out on three parallel paths: perform an applicable documents analysis, create an initial program infrastructure in anticipation of the win, and prepare the proposal itself. Some of this work may have to be performed in order to reach a sound decision on the question of whether or not to bid.

We will study applicable documents analysis later in paragraphs 10.4 and 10.5, but the central problem in this activity is to match our methods of doing business with the customer's requirements expressed in documents the customer is familiar and comfortable with (commonly military standards and specifications with DoD). In paragraph 10.4 we will also explore more fully the connection between tailoring (adjusting content) of customer compliance documents and continuous improvement of our process on the road to world class capability.

For those not familiar with DoD proposal evaluation language, BAFO means best and final offer which involves some kind of cost adjustment, generally downward. Some customer agencies and some contracts require an oral presentation in person or by video tape subsequent to proposal submission. And THE WAIT is the author's term for the period between submission of the proposal and selection when there is little the contractor can legally do to influence the outcome other than answer questions, submit a BAFO, and present an oral presentation. All three of these tasks entail high energy integration work because they do offer the contractor the only window of communication to the customer and the time constraints are very confining. The questions asked suggest customer concerns that you can now answer in voluminous detail that could not be included in the original proposal due to page limitations. BAFO offers one last attempt to fashion a winning cost figure based on new information provided by questions asked by the customer and the answers provided by all of your competitors. And the oral presentation provides an opportunity to give the customer a very brief and crystal clear message why you are the best choice.

We are primarily interested in tasks D6, D7, and DE in this book and their connection to continuous process improvement of our enterprise. The task numbers used in the blocks of these three diagrams are, like those found in many other process diagrams in the book, the authors way of coding all tasks in a computer database to explore useful information relationship concepts.

Figure 10-4 expands task D6, Create Program Infrastructure. We could simply prepare a proposal and hope for the best. However, we will help our

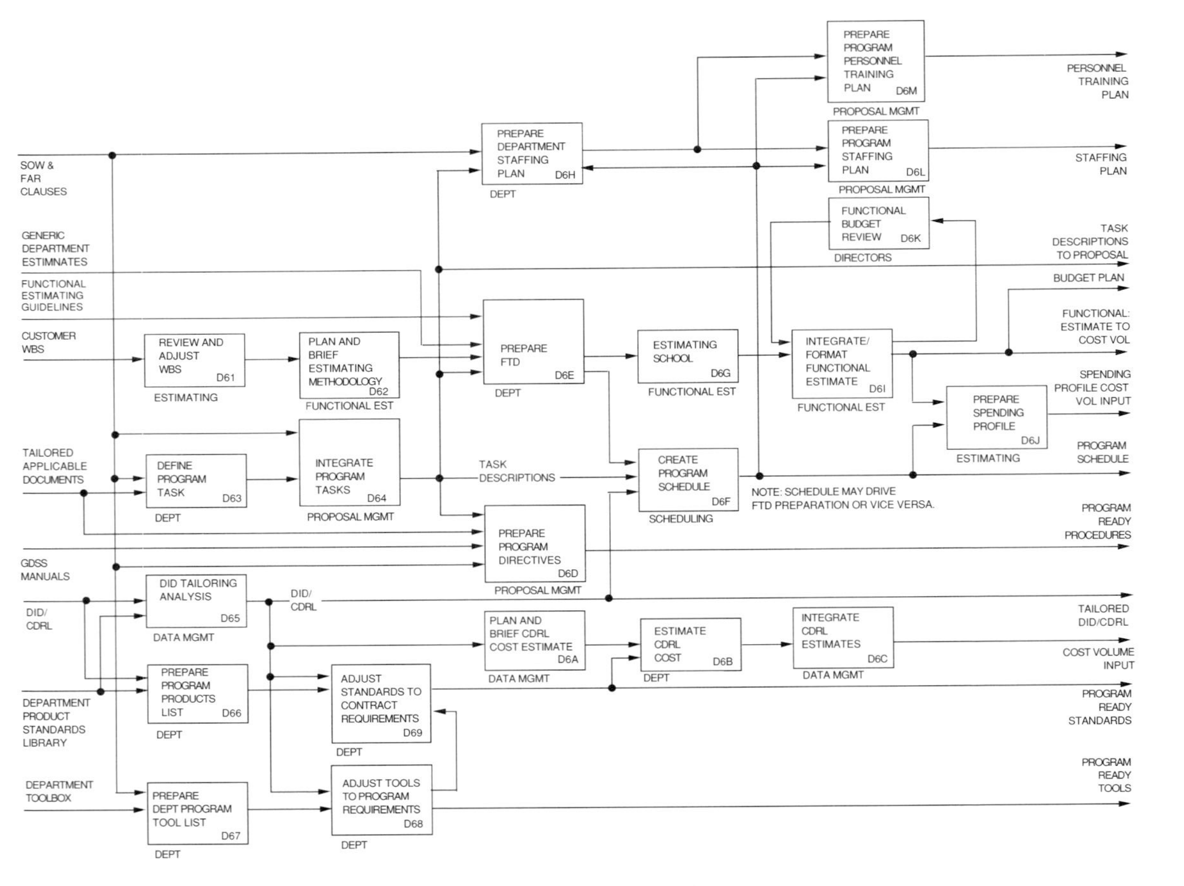

Figure 10-4 Create program infrastructure.

chances for overall success if we also take the opportunity presented by the RFP to create the infrastructure we will need in order to perform in accordance with our proposal after we are selected as the winner. Generally, this will not cost very much money above and beyond proposal costs if we are clever in taking advantage of our continuous process improvement work and work done and materials prepared for this specific proposal. The reader should have little trouble following the several paths indicated in Figure 10-3 toward the products indicated on the right margin. Some of these products feed proposal needs and others are solely related to post-RFP win readiness.

Note the generic inputs in the left margin of Figure 10-4. Each department has a toolbox they bring to the proposal effort (generally computer tools) and a product standards library containing boilerplates for the documentation products they are commonly required to produce for customers within their customer base. During the proposal period, each department may have to adjust these tools and standards for peculiar customer requirements but they should be trying to minimize the need by adjusting customer expectations through contract data item descriptions (DID) and applicable document tailoring for equivalence to internal practices.

The customer may provide a partial or complete statement of work (SOW) and work breakdown structure (WBS), but, as discussed in Chapter 6, we should study how far we can go in adjusting those structures to our planned development effort, practices, plant, and personnel base. The requirements in the system specification should drive the WBS product component and the SOW should be crafted to identify work needed in the context of our company.

The WBS is used to organize and collect the cost estimate. The SOW feeds all other program planning activity including selection of generic tasks defined by our functional organizations through continuous process improvement, linked to our training program, internal practices manuals, and department toolboxes. The integrated management plan (IMP), program plan, or system engineering management plan (SEMP) captures the detailed planning required to integrate the management of all tasks that must be performed to satisfy SOW requirements. The events and work defined in the WBS, SOW, and plan are combined with time to produce the integrated program schedule or integrated master schedule (IMS).

Task D7, Prepare Proposal, is expanded on Figure 10-5 primarily for completeness in discussing this process. Clearly the proposal volume must be consistent with the infrastructure development discussed above and this is a difficult program integration task involving most every kind of integration work characterized in Chapter 4.

10.4 Tuning our process to customer needs

Applicable documents are specifications, standards and other documents referred to in specifications, statements of work, standards, and other documents as an economical way of importing large tracts of requirements or

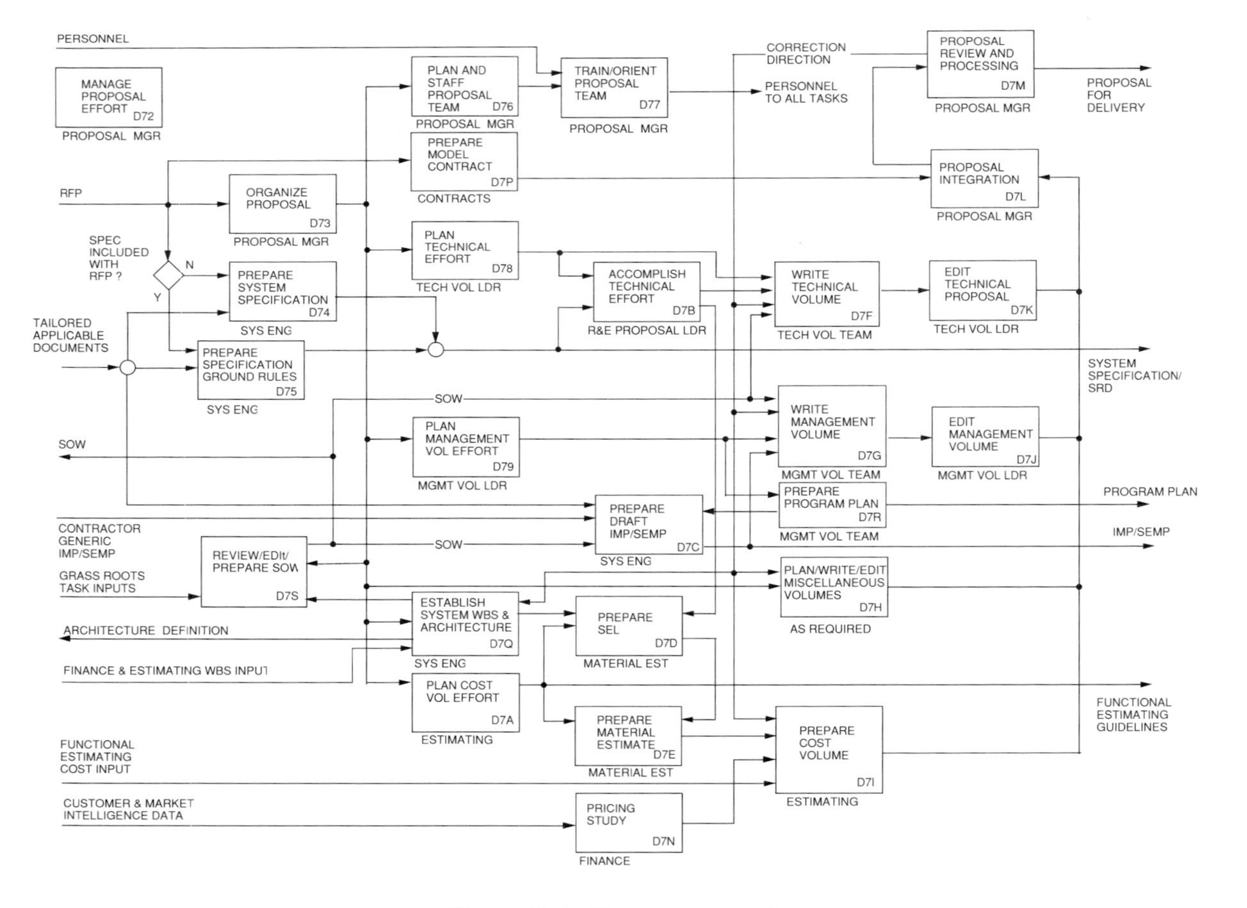

Figure 10-5 Prepare proposal.

other data without the need to duplicate the detailed contents of them. These documents provide standard solutions to problems that people have found troublesome in the past. They describe what have become standard solutions to certain problems in manufacturing, engineering, logistics, material, and many other fields. Reference to an applicable document in a calling document, such as a system specification or statement of work, applies it for the purpose and scope defined in the calling document.

The Department of Defense (DoD), as well as other customers for large scale systems, has recognized that if it wants to encourage improved productivity and quality in American industry, it must encourage contractors to develop their own procedures that are tuned to their own particular situations. This is a reversal of past policies that earlier led to contracts calling for compliance with many applicable documents offering guidance on both product design and process controls. Many companies have fallen into the habit of re-designing their internal workings for each new contract in order to comply with these many applicable documents, whether it leads to optimum customer value and contractor efficiency or not.

Some companies have independently concluded, or been encouraged by Government interest in streamlining and Total Quality Management (TQM) efforts, to recognize that they have to develop their own identity defined by their own internal facilities, personnel, methods, history, and procedures. At the same time, there is a great deal of wisdom captured within many of the 50,000 odd applicable documents, that can usefully be applied to specific situations. As a result, DoD program offices will for many years to come find a great deal of comfort in appeals to some of these documents and rightly so. Obviously, there is a condition of balance between these extremes that we need to seek out.

It will likely require several years to fully implement the Government's sincere interest in streamlining the use of applicable documents on programs. Contractors can encourage Government success in this transformation toward a better condition of balance by development of quality internal procedures that are well matched to their situation and then mapping them to the Government applicable documents most commonly applied by their customer base to programs in their product line. In the process, they will find that this mechanism also provides them with a super highway to a condition of world class capability in their product line through continuous process improvement.

The general pattern involves development of a set of internal practices for each process or task you find it necessary to apply on programs. Then, you develop a list of applicable documents your customer base commonly applies to contracts or defines their value system in accordance with. We then map these documents to our internal practices and tailor them to agree perfectly with our internal practices. Tailoring is an editing process to change the content of a document. It can either be done using a legalistic or in-context method.

In the legalistic tailoring method, we make a list of changes to the document such as:

(1) Delete paragraph 3.2.3.5
(2) Change paragraph 3.4 to read: "3.4 The pressure vessel design shall be proof tested for 2.5 times their normal maximum operating pressure."
(3) Add paragraph 3.6.5 as follow: "3.6.5 Limit the pre-shipment test profile for a production pressure vessel manufactured in accordance with approved procedures to 3 excursions to a pressure level no more than 2 times normal operating pressure for 15 minutes each."

In the in-context method, an electronic copy of the document is marked up with a word processor to line out deleted text and highlight added text. While this method is much easier to use because you can see the context of changes, very few of these kinds of documents are available in electronic media unless you scan and edit them locally.

10.5 Applicable documents analysis

When we bid on a contract for which the customer has offered a particular set of applicable documents, we feed back to the customer, to the maximum extent possible, our tailoring that makes their documents equivalent to our internal practices documentation. The customer can manage the program in a familiar context and our employees can follow the same set of internal practices they have been trained on and become experienced with. There was a time when this action would be referred to in DoD as non-responsive or arrogant, but DoD has begun to understand the need for contractors to perfect their own processes.

That is the advantage of an internally defined identity in terms of internal practices, it allows the employees to always follow the same process becoming experienced in that process. It supports an internal training program focused on the generic processes defined in company practices that become the textbooks for that training program. It supports the practice-practice-practice notion thorough which world class athletes become that way. It also will provide the customer with the best possible value that your combination of plant and personnel will permit. These advantages will be short-lived unless you also implement a continuous process improvement program as suggested earlier in this chapter. We have expressed many ideas in this chapter that you may have thought of as separate entities previously. It requires effective cross-process and cross-function integration to mold these entities into a dynamic whole.

Figure 10-6 offers a process for structured tailoring that fits into the RFP-Proposal diagram in Figure 10-3. This process is repeated for each proposal. From these tailoring exercises you will from time-to-time gain insights into how to improve your internal practices. Likewise, during actual program performance, you will find better ways to do things than described in your internal practices. These are lessons-learned signals that should stimulate generic process improvements available to all new programs.

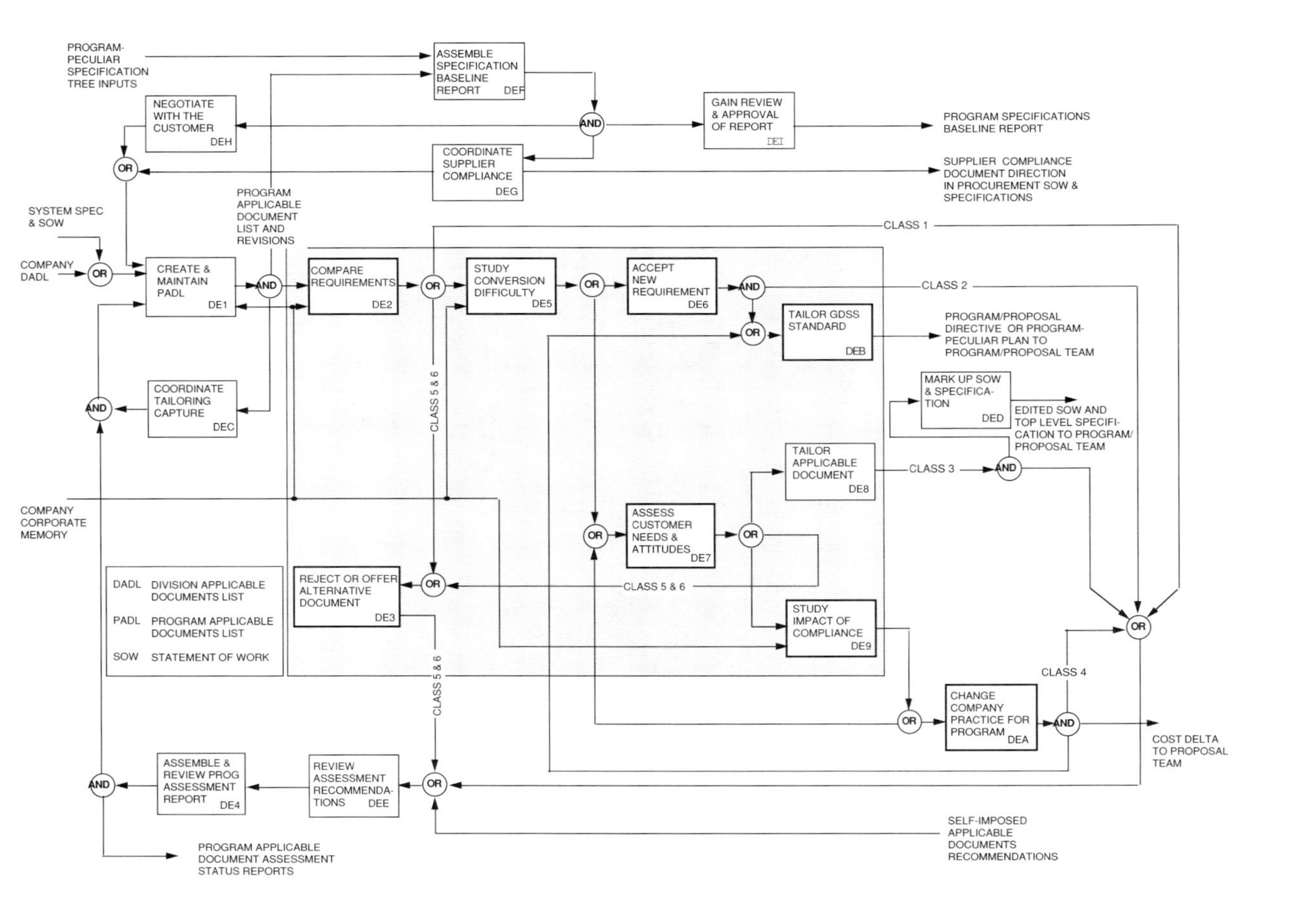

Figure 10-6 Structured applicable document analysis.

10.6 Program audits by functional departments

Given that our functional departments have developed internal practices to guide our personnel in performance of program tasks, it will be helpful to both program and functional management for functional management to audit program performance to those practices. Functional management is thus able to find out what is going on and to provide a useful service to programs. The audit should be conducted to find out the degree of faithfulness in the program application of a given practice using the functional practice description as the baseline.

The audit could be carried out in a very subjective fashion where the functional manager simply gives a pass/fail or numerical grade in a 0 to 10 or 0 to 100 range. A better way to do it would be to ask each functional department to prepare an audit checklist for each of their practice descriptions. The answers can then be used to assign an objective score that can be tracked over time providing a useful metric.

If your company allows programs to tailor or adjust functional practices for programs then this audit may have to be made based on a program manual description of the differences. If your company allows programs to develop their own practices independently of the functional departments then auditing will be money wasted and it will take your company a very long time to be counted among the great system houses. You will not be able to take advantage of the practice-practice-practice and continuous process improvement notions encouraged here.

10.7 Benchmarking

The final continuous process improvement process offered is benchmarking. If we only rely on our own experience in our search for excellence, there is a danger of the development of an incestuous group-think mentality that will shield your organization from thinking new thoughts and seeing the value of potentially valuable alternatives. We need to assure that we have access to ideas from outside our immediate organization and we need to consciously recognize the need to compare our performance against these ideas. We should only seek out what we consider to be excellent examples with which to compare our performance in the form of other companies we respect, especially if they are our competitors. This is called benchmarking.

The unfortunate thing is that this kind of information is not always easy to obtain. Your competitors will be reluctant to allow you direct access to it. But, it is not really necessary to engage in covert industrial intelligence operations to find these things out. One source in a large multi-division organization is your other divisions. But, a source that is open to all is provided by the many trade and professional organizations that touch on practically every facet of your business. Examples are: the National Council on Systems Engineering (NCOSE), American Institute of Aeronautics and Astronautics (AIAA), Society of Automotive Engineers (SAE), and Institute of Electrical and Electronic Engineers (IEEE). There are members of these organizations that come from all of the best and most successful companies in your field. These members will offer papers covering their best ideas most often derived from their experience in those companies. Personal interaction between your employees and other members will also provide additional insights into alternative approaches of potential use to your company.

The functional audit process and benchmarking can be combined to produce a two-figure metric for your processes. The first figure could be derived as noted under functional audits. The second figure would be derived by an audit of your in-house practice against a benchmarking standard. This pair of figures gives your management added information about how best to apply scarce resources to improve your company's performance.

10.8 Where is your process description?

You are encouraged to seek out your company's written description of your system engineering process. Does it exist? If it exists, is it up to date and being followed? You will likely find that the answers to these questions is no. In the book *System Engineering Planning and Enterprise Identity* you will find a generic system engineering manual (SEM)/system engineering management plan (SEMP) discussed at some length in Chapter 6 of this book. This document may offer you a beginning toward the goal of defining your system engineering process. Once defined, you will find a need to continually improve it as discussed in this chapter based on your experiences on programs.

Part III

Product integration

chapter eleven

Architecture definition

The work described in this chapter is properly accomplished early in a program to define and organize the product system and associate the development organization with responsibilities for development of product elements. There is a strong flavor of process integration embedded in architecture definition, as you will see, but the principal focus is on determining what shall be in the system in terms of hardware, software, people, and facilities. It is an example of cross function and product integration (integration type 20 from Table 4-1). Specialists from many functional organizations must cooperate in the task to develop the complete architecture of the system and the several different views of the system we will overlay upon the architecture.

The word architecture is used in this book to mean a hierarchical relationship between everything that is in the system. The very top of the hierarchy is the system itself. The next layer down will consist of some number of major elements in the system. Each of these is further broken down into their components and this pattern continued to the lowest tiers of the system in each branch. When we begin to develop a new and unprecedented system, using the top down structured approach recommended in this book, the only element of this hierarchy we know is the system. During the early development phases we need to completely define the architecture by some organized means based on what we know the system must do.

If the system has been previously developed and our purpose is to significantly modify or re-engineer it, our course should be a little different in the beginning, but the same end game is valid for those parts of the system that must be changed. We must understand what the new need (goal or purpose) is and then determine what in the existing system architecture remains useful in satisfying that need. New or significantly changed functionality that cannot be satisfied by existing or modified resources will have to be satisfied by new resources that we can gain insight into from the methods discussed in this chapter. If, in the unlikely event the original analytical data is available for the system, we can begin by determining impacts based on the changes in the customer's need and then ripple those changes down through the analysis to find architecture impacts. In the more likely case, we will have no information from a prior system development cycle for the existing system even if an organized systems approach had been applied during its development. In this case, we should quickly recreate the analytical basis for the system using methods discussed in this chapter being alert to shut off analysis on strings where we conclude no changes should be made.

One computer software development model called for the analyst to understand the existing physical model and then the existing functional model as a basis for then creating a functional and physical model for the modified system. One criticism of this approach was that no customer could restrain their interest in the new development long enough for the analyst to understand the old system from which they were trying to escape.

This is a valid concern that the development team must be conscious of. Any attempt to recreate a proper functional system description should be carefully monitored against predefined goals for which there is an iron clad definition of completeness criteria.

11.1 Structured analysis

Prior to discussing architecture definition, let us first introduce a rigorous top-down method for determining an appropriate architecture for a system. For more expansive coverage of this matter, refer to the book *System Requirements Analysis.* The structured analysis method common to the system requirements analysis process developed by several major Department of Defense weapons system program offices and aerospace contractors involves some kind of diagrammatic aid to understand needed system functionality followed by assignment or allocation of that functionality to things, elements, components, or items in the system. This approach can be applied effectively to any kind of system whether for commercial sale or DoD contract. Figure 11-1 illustrates this whole structured decomposition process.

Perhaps the most traditional approach, where the whole system is not computer software, is to use a functional flow diagram as a tool to explore what exactly the system must do. As the analyst gains understanding about needed system functions, this functionality is allocated (assigned) to particular system elements responsible for satisfying that functionality. Where some doubt exists about the most appropriate allocation, a trade study is performed to clearly reveal the best approach of two or more possible ones (hardware versus software implementation, for example). Please note that we are using the word function to denote an activity or capability of a system. Previously, we have used the word function to refer to the organizational elements of a matrix-managed company. As an alternative to functional flow analysis, we could use one of several software development techniques and tools (if the item is software), hierarchical functional analysis, behavior diagramming (as used with the Ascent Logic tool RDD-100), or variations of these diagrammatic approaches to get insight into what exactly the system must do.

The functional allocations that result from this process become the basis for both the identification of the system architecture, through allocation, and

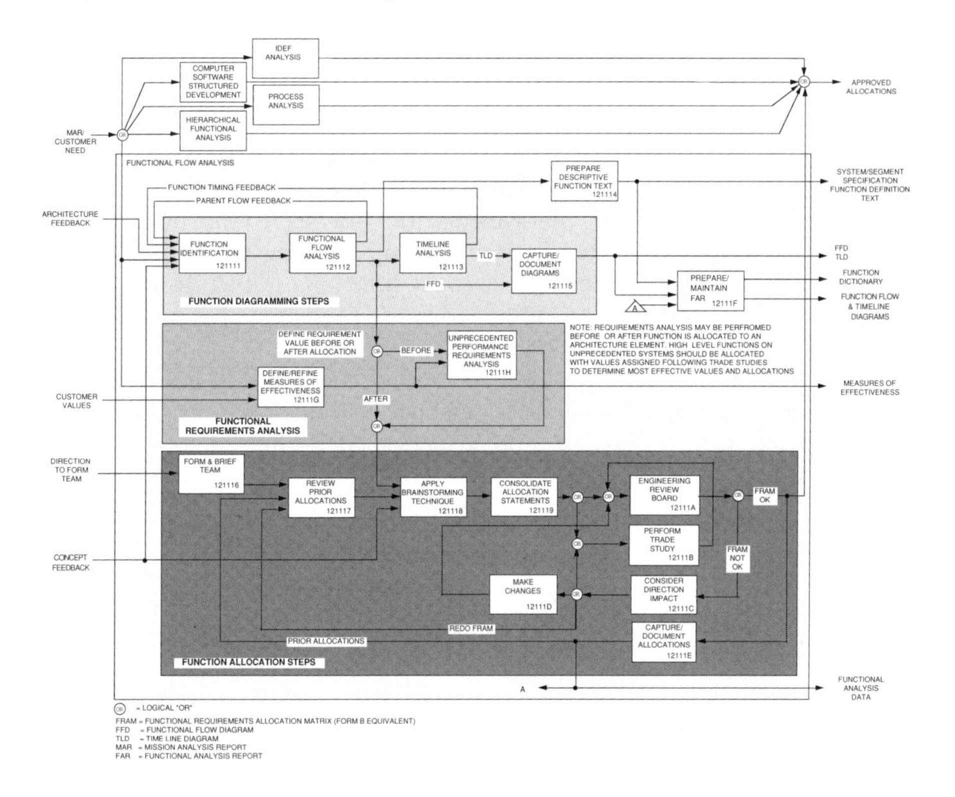

Figure 11-1 Structured decomposition process.

precursors of performance requirements appropriate to the elements of that architecture. This is the heart of the structured development process. The act of allocating a function to a system element places a demand on the development team to transform that allocation into a performance requirement for the element to which the function was allocated. The brief text of the function statement becomes a primitive form of the final performance requirement.

Many system engineers experienced in the development of unprecedented systems, like Mr. Archie Vickers of PRC, would insist that the function statements be converted into performance requirements prior to allocation as a way to avoid a leap to point designs. This alternative offers advantages particularly for unprecedented systems.

The resultant system elements are assembled into a hierarchical block diagram, the top item of which is the complete system. The method for assembling the evolving architecture into related family trees suggested here is to group them based on minimizing the cross-organizational interface relationships which are defined as interfaces between product elements where the organizations responsible for designing the terminal elements are different. It is exactly these kinds of interfaces that will result in the principal system development difficulty, so we can reduce the development difficulty by reducing these kinds of interfaces. We cannot eliminate them since they are responsible for the richness of the system, but those that remain should be very clearly known to the development team and responsibility very clearly established.

In Figure 11-1, you will note that the functional flow diagramming approach is expanded in detail. Three major steps are noted, each further expanded on the diagram: (1) first we must identify needed functionality and link it into a sequence of blocks representing the functionality; (2) if we choose to do so, we may expand defined functions into fully quantified performance requirements; and (3) the identified functionality must be allocated to things in the physical model, the architecture.

Four alternative decomposition approaches are noted in the figure but not expanded. Note that this process must be cyclically applied to lower and lower levels to fully define the problem space. As we allocate functionality to the solution space, the physical model, the architecture, the concepts developed for those items should be fed back to the decomposition process to refine the lower tiers of the functional analysis with higher level solutions. This acts to tune the lower tier functionality to the evolving solution and greatly speed up the process. If done too zealously, this can also encourage a rush to point designs of the past. Pacing this process is an art form learned by doing.

When a function is allocated to computer software, the team responsible for developing that software will prefer to apply one of the many computer aided software engineering (CASE) approaches and tool sets to the further analysis of needed functionality. That should, of course be permitted. In fact, while some system engineers will be shocked, the author believes that a team should be permitted to use any of the noted decomposition techniques based on the kind of product and the experience of the team members. As a result, the complete decomposition analysis may be composed of a mixture of techniques.

In this chapter we will explore only how to establish the system architecture and map it to other hierarchical structures of interest to the development team. In Chapter 12, we will pick up on the interface analysis process that is useful in refining the architecture along the lines of minimized cross-organizational interfaces.

11.2 Architecture synthesis overview

One of the principal products of architecture synthesis is the architecture block diagram, an example of which is illustrated in Figure 11-2. The top item is the complete system. This block can be allocated mindlessly from the top function, the customer's need statement. The next tier includes major elements of the total system. This pattern keeps breaking down to the lower levels of system composition. As noted, structured analysis generates allocations of needed system functionality to specific physical entities through trade studies, an appeal to historical precedent, good engineering judgment, and respect for customer direction on use of specific resources in the system. Customer furnished property is an example of the latter and this occurs when

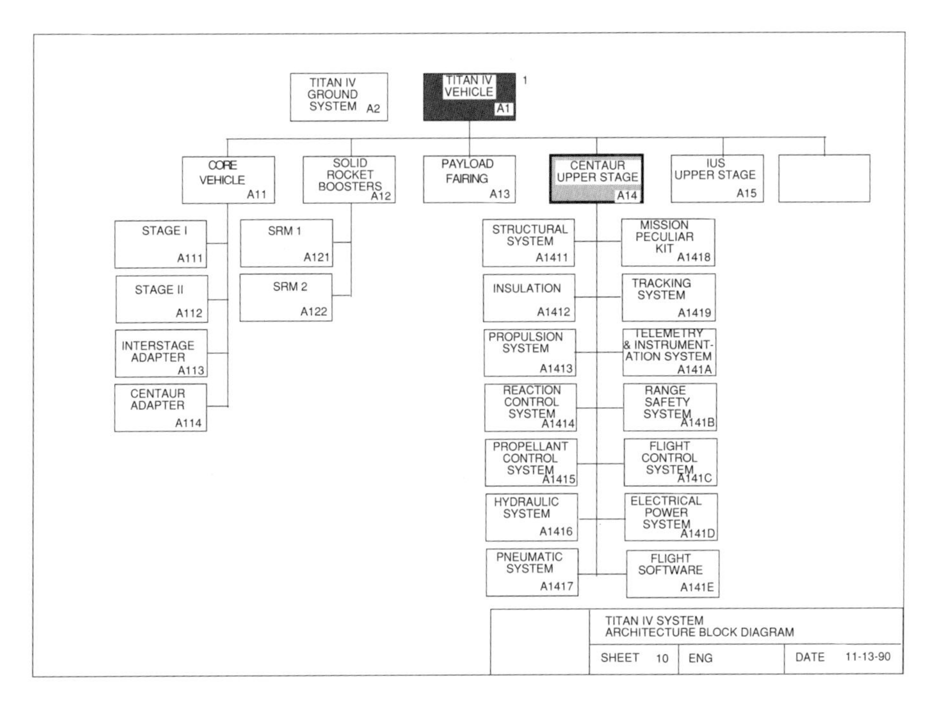

Figure 11-2 Typical architecture block diagram.

a customer has residual property from other systems they wish to make continued use of in a new or updated system.

On any given project, a Program Integration Team (PIT) should be made responsible for the complete architecture diagram. The PIT should conduct the structured decomposition process down to the point where it yields allocations to elements that can be assigned to product-oriented integrated Product Development Teams (PDT). Once that assignment has been made, the responsible PDT should continue the structured decomposition process within the confines of the element(s) allocated to them.

The engineering community, in concert with the production and logistics leadership on the PIT and PDT must decide in team efforts how to organize these allocations into families of things. They must map the evolving architecture to: (1) organizations responsible for the development of each item (thereby giving birth to PDTs), (2) the planned manufacturing process flow, (3) the customer's interests in managing the development process through configuration items, (4) make or buy considerations, (5) specification needs, and (6) the customer's work breakdown structure (WBS) that will be used to organize and report contractor cost and schedule performance. All of these views of the system should have an influence on the final architecture synthesis.

Mistakes can be made in the process of allocating functionality to architecture. The system engineering component of PIT must continually evaluate the evolving architecture for suboptimum structure and interface relationships. Suboptimal structures will commonly result in reduced product system efficiency and development difficulties. Chapters 8 and 14 discuss this problem in more detail. In that most system development problems occur at the interfaces where functionality differences coincide with organizational responsibility differences. Optimum interface is defined as that interface condition characterized by a minimized interface count between elements under development responsibility by different design agents (different teams or contractors). Chapter 12 covers this area of cross organizational interface in detail.

The results of the interface analysis must be fed back into the architecture definition process to close the loop on this system optimizing activity. In the process you may find many other changes will be triggered, but this is part of the iteration, or churning activity, that must occur as we become smarter about the system we are creating. We proceed from the simple to the complex, from the general to the specific. On the route we may find that ideas we had when we were less knowledgeable were not the best ones. We must have the courage in this event to change those ideas as early as we possibly can since it is much less costly to do so early in a program than later.

The same people on the PIT responsible for functional allocation at the system level should also be responsible for architecture synthesis and system interface optimization. These three activities are closely related and offer synergistic and serendipitous opportunities where the same people are doing all three. If people with different reporting paths are responsible for parts of this work, they must be collocated physically and encouraged to interact to the degree that they effectively form one team.

Figure 11-3 illustrates the process we will be discussing in this chapter. The discussion opens with architecture block diagramming and follows up with mapping of this architecture to several important hierarchical views of the system including: work breakdown structure (WBS), manufacturing breakdown structure (MBS), drawing breakdown structure (DBS), customer configuration items list, and specification tree. Configuration items are identified by the DoD customer as the major elements through which a DoD customer would prefer to manage a development program. Other customers may refer to them as principal elements, end items, or contract end items (NASA). A commercial products organization is, of course, unconstrained in deciding through which items it shall manage the evolving program but should pick some that allow them to focus on managing the important, driving aspects of the evolving system. During early program phases the customer may ask the contractor to supply a candidate list of configuration items. We will, later in this chapter, describe a process for assembling that list

The interface analysis block on Figure 11-3 will be covered in Chapter 12. But, note that we are suggesting that as the architecture unfolds, it should be reviewed (by the PIT) for interface appropriateness. This is the activity mentioned above where we are trying to minimize the number of cross-organizational interfaces. The product of the architecture synthesis process can be captured into a single baseline definition document for record purposes as illustrated on Figure 11-3.

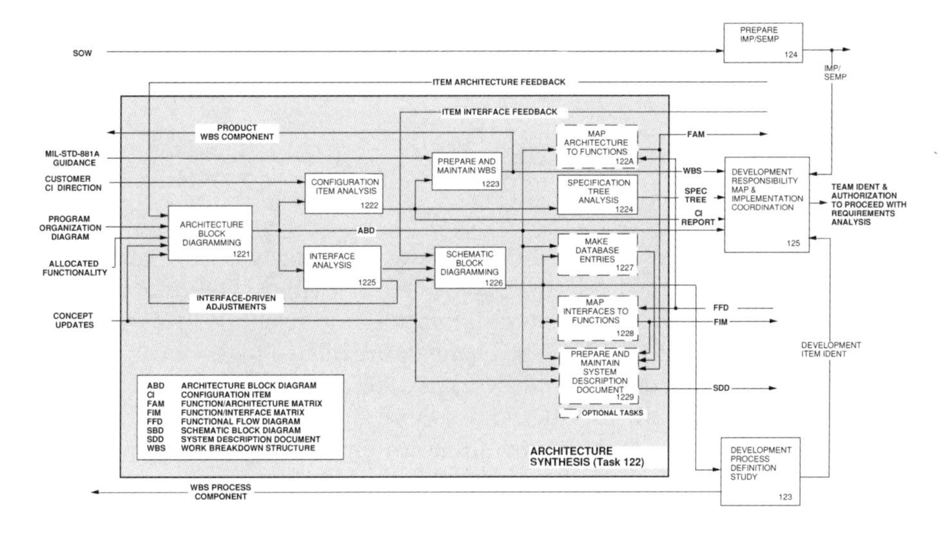

Figure 11-3 Architecture synthesis process.

11.3 *Architecture block diagramming*

An architecture block diagram is a simple hierarchical block diagram consisting of blocks that represent the things composing the system and interconnecting lines that show parent and child relationships. Commonly one item is composed of two or more subordinate items and this pattern is repeated all the way from the system block to the lower tier composed of things that will surrender to detailed design by small teams or procurement at that level.

In placing the things on the diagram, identified through functional decomposition, we must consider many views and identify diagram overlays based on these different views as we proceed. The next section describes some of these different views that must be evaluated concurrently as we try to reach a conclusion on the best system organization. These different views will not always be in agreement on the best architecture arrangement so the team must provide for a decision-making process that may entail trade studies where very difficult decisions are involved.

The PIT should be responsible for top level architecture definition and this responsibility should be passed on to the PDTs for items for which those teams are responsible. PIT should monitor lower tier architecture development and integrate those lower tier components into the system architecture block diagram.

Figure 11-2 uses a base 60 architecture ID notation to avoid use of decimal points between levels. The decimal delimited codes have the advantage of handling many items at any one level but the decimal points do not sort well in simple database systems some people may be forced to use. The base 60 system uses the 10 Arabic numerals plus 25 capitol English letters (capital "O" not used to avoid confusion with the numeral zero) and 25 lower case English letters (lower case "l" not used to avoid confusion with the numeral one). This system permits identification of up to 60 items at any one level which will accommodate most system needs.

11.4 Architecture overlays

11.4.1 WBS overlay

The work breakdown structure (WBS) is a cost-oriented view of system structure defined for the Department of Defense in MIL-STD-881. In Chapter 6 we saw how customers manage program cost and schedule by associating all program work with a WBS. WBS elements are related to those responsible for performing the work through work authorization numbers and plans. The WBS includes both product and service elements. The only portion, however, of that tree that we are immediately interested in here is the product portion. Figure 11-4 illustrates a typical WBS.

A complete WBS may include multiple structures differentiated by a prefix code. We might have one prefix for recurring work and another for non-recurring, for example. One prefix might be established for the non-recurring design activity and another for non-recurring flight test activity. The prefix codes essentially establish a layered structure such that a given WBS number may appear in several prefixes. Cost can be collected in two axes this way. For example a WBS 1300 might be for the structure of a launch vehicle. All cost corresponding to development of the design for the structure might be collected in WBS 123-1300, where 123 is a prefix for non-recurring design development work. The work corresponding to structure testing might be collected in WBS 342-1300, where 342 is a prefix for non-recurring test and evaluation. All work and cost related to the structure can be compiled by adding all of the WBS 1300 figures from all of the prefixes. At the same time, all non-recurring work can be accumulated by adding the total of all non-recurring prefixes.

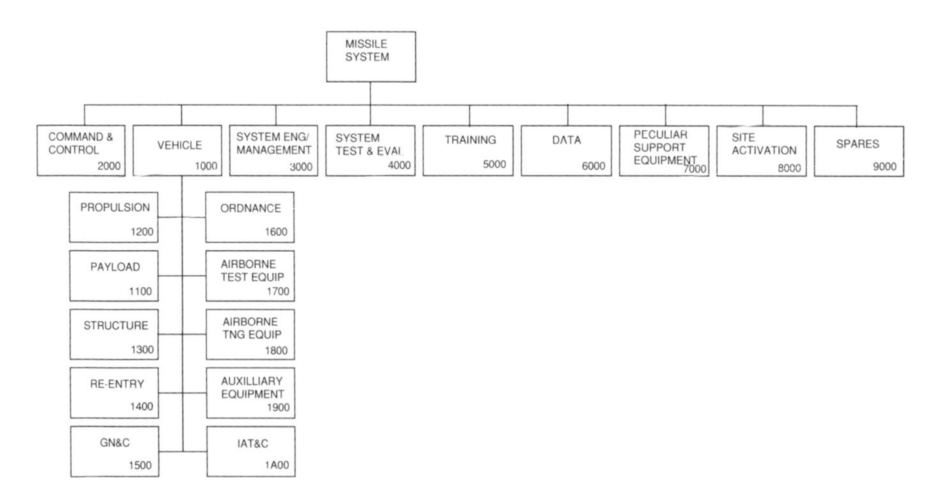

Figure 11-4 Typical WBS diagram.

One of the most contentious problems in the early phases of system development is the imposition of a cost-oriented structure of the system onto the engineering organization that is trying to understand in an orderly way what the system should consist of. It is all very well for the finance world to develop an initial work breakdown structure (WBS), but they should recognize that the structure of the system should be defined in terms of its functionality not some arbitrary finance standard for which the finance computers happen to be programmed. The WBS should be adjusted concurrently based on the results of the architecture block diagramming process jointly by finance and the engineering community through membership on the PIT.

The reason that this is not always possible is that customer and contractor finance groups generally have access to cost models that are insufficiently flexible to change and that the inflexibility of the models is matched or exceeded by the people who staff these groups. This problem is an example of past foolishness in allowing functional organizations to build walls between themselves rather than cooperating in a teamwork environment.

The engineering community should, none the less, assertively press its case for a functionally organized view of the system architecture and make an effort to force adjustments in the WBS to reflect this structure. If it is not possible to move the heavy weight of an inflexible WBS, be prepared to simply map the architecture to the WBS in the architecture dictionary. The Finance and Program Office people are right that everything in the system must belong to some WBS in order to properly manage program cost and schedule.

MIL-STD-881 includes several appendices that give suggestions for a WBS arrangement for different kinds of systems. These include deliverable product system elements as well as contractor service elements. The WBS seeks to embrace every possible program cost element. It is best if the WBS can do this so as to reflect the physical and functional organization of the system. Better still if the WBS can be used directly as the Architecture ID system. One way to facilitate this result is to overlay the ABD with the WBS assignments. Whenever you observe a crossing of the lines between the physical, functional, and cost (WBS) organizations of the system, you have a system organizational conflict. What can happen if this situation prevails?

A common problem in system development is a discontinuity between the functional and physical organization of the system. If you follow the pattern expressed in this book, the system will be decomposed into functional subsystems each the responsibility of a design group. Let us say by way of example that the WBS for a system matches the functional breakdown because it is clear that the functional subsystems line up with the way the contractor is organized and therefore the WBS will mate up well with reports generated by the contractor's cost schedule control system.

These functional subsystems must be integrated into one or more physical entities. If the design evolves to allow one or more subsystems to cross over the boundaries of system elements that form obviously separate physical components that are respected in the engineering drawing tree and manufacturing plans, it can result in great difficulty in determining the cost of the physical entities. If one or more of the physical entities is a configuration item (see next section) through which the customer manages the program, the difficulty can be increased for the customer.

For example, picture a space transport launch vehicle upper stage that is composed of two major physical end items, a main body and an equipment section. Now lay on eight functional subsystems with two of them wholly contained on the main body section, three wholly contained on the equipment section, and the remaining three including components spread between the two items. If the WBS is organized about the functional structure of the system, it will be very difficult to determine the cost of either one of these physical entities. It is important to avoid crossing of these physical, functional, and cost boundaries in elements that have a useful separate physical existence.

Space transport launch facilities, oil refineries, and computer network installations provide other examples of this problem. These complexes may consist of facilities interwoven with functional systems for fluid handling, electrical control, and communications. The best way to avoid conflict in these cases is to recognize that the customer is taking delivery of two kinds of things: (1) facilities and (2) operating systems that are installed within these facilities in some pattern. The WBS can reflect these two sets with little difficulty even when the functional systems cross facility boundaries. In this case, the complete complex is an integrated whole that will always be used together.

11.4.2 Configuration item overlay

Customers like DoD and NASA will wish to especially identify certain items through which the complete program will be managed. In DoD these items are called configuration items and in NASA they are referred to as end items. This intent is to select items in a band across the architecture such that every thing in the system is contained in one of the configuration items. Figure 11-5 illustrates this concept. The items corresponding to the shaded portions would be the subject of major reviews held by the customer that occur at major program milestones. The PDT structure should be coordinated with this arrangement to simplify and avoid program control problems.

The customer may select high level items where they wish to manage the program at a high level or detailed items where there are many new development items and they have the program management resources to cover the details. A very complex system involving new technology may have up to three layers of configuration items. Refer to *System Requirements Analysis* for detailed coverage of a comprehensive item selection logic.

11.4.3 Specification tree overlay

The specification tree defines the elements in the ABD that require formal documentation of their requirements in a customer specification format,

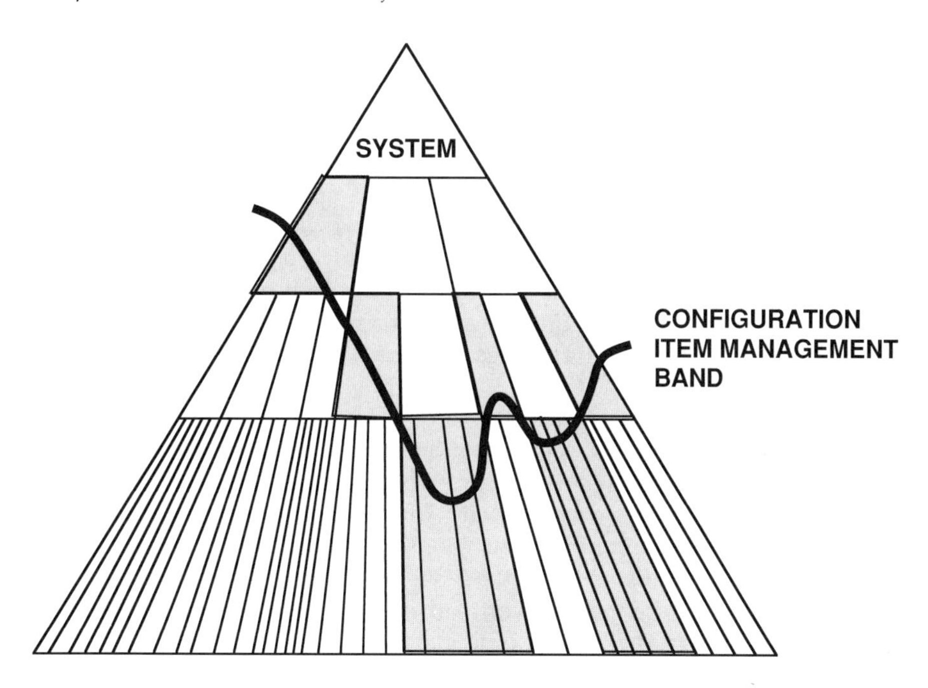

Figure 11-5 Configuration item management band.

procurement specification format, or in-house format. The identification of configuration items carries with it the need to write specifications which will be contractually binding and require formal verification of product compliance with them. Specifications are the repository for the output of the requirements analysis process for a given item. The requirements content of specifications is what holds the whole structured development process together. They ensure that the solutions to the many small problems into which we decompose the ultimate function, the customer's need, will fit together to solve the system problem. This is attained through traceability of requirements and the specification tree indicates the principal traceability paths that must exist.

The tree can also be annotated with specification type information such as the various configuration item types, procurement specifications, and in-house types. The specification tree defines the elements in the ABD that require formal documentation of their requirements in a customer specification format, procurement specification format, or in-house format. The identification of configuration items carries with it the need to write specifications which will be contractually binding and require formal verification of product compliance with them.

The specification tree will show three different kinds of specifications. First, and commonly at the top of the architecture, we will find the customer specifications (system and configuration or end item specifications). We will also wish to clearly define the requirements for any items that we will procure where those items will require some development and testing. These requirements should be captured in procurement specifications. Finally, there will be some items that we will accept responsibility for development of and also conclude that we should define their requirements prior to performing design work. These we might call in-house requirements documents.

You may have to have two versions of a specification tree. A DoD customer is mainly interested in the tree showing only system, segment, and configuration item specifications. The contractor may want an augmented tree with the other specification types indicated as well. In any case, the specification tree should be an overlay of the architecture diagram. There should be nothing on the specification that is not on the architecture diagram.

Figure 11-6 illustrates a typical diagram. In this example the blocks are shaded to tell what kind of document and the block contains the specification number in the lower right corner of each block. It is useful to show the top level specification tree in diagrammatic form, especially down through the level where all customer specifications are included. At lower levels it is better to use a tabular list to avoid the cost of maintaining the graphics. If you have an architecture database, you can include specification numbers in another field and arrange for an indentured output report.

11.4.4 *Manufacturing, procurement, and engineering breakdown structure overlays*

The manufacturing, procurement, and drawing breakdown structure activities are not illustrated on Figure 11-3 but they feed off from the architecture block diagramming activity. In the manufacturing and procurement cases, we seek to influence the organization of the system into major components that neatly align with our manufacturing plant and capabilities and our procurement plans.

Manufacturing will wish to be able to identify the assemblies that are worked upon at its work stations in terms of specific engineering drawing numbers in the simplest possible way. We want our production line to have to deal only with physically coherent wholes where possible and avoid situations where a manufacturing work station has to work with items only partially covered on two or more drawings. Ideally, the engineering drawings should map to the work station activities view of the product in a simple one-to-one relationship. The more complex the relationship between these two views of the architecture, the more difficult the integration task during production.

The PIT manufacturing engineer can achieve this desired simple relationship by planning the manufacturing facilitization concurrently with product design development and influencing the system architecture structuring accordingly concurrently with engineering efforts to map the architecture to the engineering drawing structure. Manufacturing membership on the PIT must define exactly how the system will be manufactured. What facilities

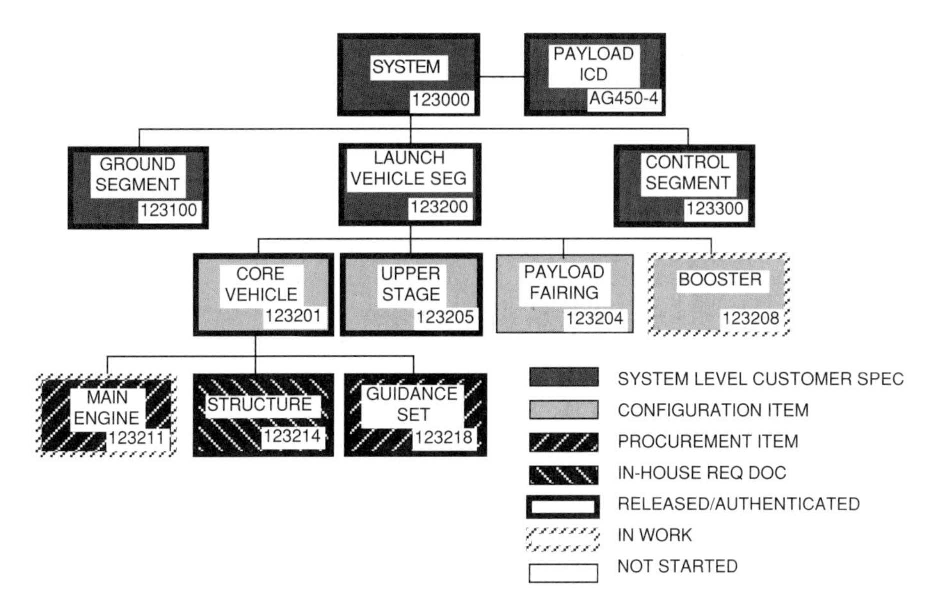

Figure 11-6 Typical specification tree.

will be used? How will these facilities be allocated to particular production processes? What work stations will be included within the process? Answers to these questions will encourage a particular physical organization of the product.

The architecture should also evolve with conscious identification of the results of the make-buy decision process. Each item to be procured should have a separate engineering drawing and appear on the architecture diagram as a single entity at some level of the architecture. When this condition exists, procurement can contract with a supplier with a very simple drawing. All of the notes and data on the drawing should apply to the procurement. When a mismatch is permitted to develop in this area, procurement must fabricate special instructions that may even countermand engineering drawing instructions. This leads to complexities in the manufacturing process because a part may not comply completely with the engineering when received from the supplier. Special manufacturing planning may be required to perform steps on purchased parts before they are installed.

The manufacturing and procurement breakdown structuring of the architecture should be one of two driving forces, the other being minimized cross-organizational interfaces. Once the architecture is screened for manufacturing breakdown, it can then be used as the basis for the engineering drawing breakdown structure (DBS). If you come out of this activity with minimized cross-organizational interface and consistency between the engineering drawing structure and the manufacturing process, you are well on the road to a money-making program that produces an item that will be highly valued by your customer.

Clearly, it requires procurement, manufacturing, and design knowledge concurrently applied in a team environment to properly structure the system architecture. Finance and configuration management also must cooperate for WBS, configuration item, and specification tree overlays. With all of these specialized groups involved, it also makes sense to insist on a system engineer to coordinate the overall architecture structuring ensuring that all of these views are respected and conflicts are resolved based on sound logic derived from the conflicting inputs from the several specialized system views. It is precisely this conversation, possibly very intense in nature, that must take place in block 1221 of Figure 11-3.

chapter twelve

Interface development

Interface development is a core system engineering activity because of the fundamental development technique of decomposing large problems into many related small ones. The small problems have interfaces between them and these interfaces must have the same solution on both terminals. The difficult part of this task is that each of the two terminals may be the responsibility of a different person or team. All architecture items should have only a single party assigned responsibility. Interfaces commonly will have two or possibly even three parties responsible for their design. Clearly, the work of these people or teams must be integrated in an application of cross product integration. This chapter was derived from an earlier work by the author titled *System Requirements Analysis* published by McGraw-Hill in 1993 and is included here with the permission of the publisher.

12.1 Interface analysis

Interface analysis is one of the methods by which we arrive at an optimum architecture for a system before we commit to its design. We define precisely where the interface planes are in the system, who is responsible for them, and ensure that the principal engineers responsible for the interface terminals and the media all have precisely the same understanding of the requirements and design concept for each one. In the process we may affect changes to the system architecture derived from allocation of system functionality and influenced by the several overlay views discussed earlier. The desired goal is to refine the match between how the design team is organized and how the product system is organized in order to simplify the system development process and assure that we have accounted for all needed system functionality.

This book encourages the use of N-Square diagramming techniques augmented by specialized analytical techniques corresponding to the several engineering disciplines (avionics, fluid dynamics, and so forth) to determine the optimum interface relations between system elements. Generally, the analyst should seek to maximize the capacity for interaction between system elements while minimizing the system need to interact. Generally, this results in minimizing the number and complexity of system interfaces.

Given that we have identified an optimum interface condition within the system through interface analysis, we need to communicate the results of that analysis very effectively to all development team personnel to ensure they are all working to this model across their own portions of the model. Schematic block diagrams and interface dictionaries can be very effective in satisfying the communications objective. Schematic block diagramming is a graphical methodology for reporting upon the preferred interfaces defined through interface analysis. An interface dictionary simply lists every interface and correlates them with other system information.

12.2 Interface defined

An interface is a plane or place at which independent systems or components thereof meet and act or communicate with each other. An interface is characterized by two terminals, each touching one element in the system architecture. An interface is completed between these terminals via an interface media such as physical contact, electrical signals in wiring, fluid flow in plumbing, or a radio signal in space. The interface media is provided by either an element of the system architecture or the system environment. The interface is not the media itself, rather the functionality facilitated by the media.

For example, as a function of being bolted together, two physically interfacing components will remain attached. As a function of being connected by a wire harness, an aircraft on-board computer may send a command signal to the aileron actuator of an aircraft that causes the aileron to move interacting with the air mass of the environment, in relative motion with respect to the aircraft, to roll and turn the aircraft (swept wing aircraft with high degree of yaw-roll coupling assumed) such that the command signal is nulled out in the on-board computer as a function of the guidance set detecting approach to the direction commanded.

It is common practice to associate an interface with one of three types: functional, physical, or environmental. A physical interface involves the form and fit of mating parts. Examples of physical interfaces include: mounting bolt patterns, drive shaft flange connections, mating wire harness connector physical attachment, and the tires of a fighter plane resting on the apron.

Some examples of functional interfaces are a 28VDC signal passing from a solenoid driver in an on-board computer to a valve solenoid in a pressurization control unit, a digital data stream flowing from an instrumentation control unit to a flight data recorder input port, and the flow of liquid oxygen from a rocket propellant system to the engine.

Environmental interfaces can exist between two items when the natural environment communicates environmental stresses between them. A reconnaissance camera company once argued that one of their cameras could take excellent pictures through the engine exhaust of an unmanned reconnaissance aircraft. If this were true it would have made it possible to outfit a very inexpensive unmanned photo reconnaissance bird. Unfortunately no one questioned the environmental interface provided by the jet engine exhaust between the desired imagery and the camera system before money had been spent to prove the existence of an irreconcilable environmental interface incompatibility.

12.3 The interface dilemma

Included in the definition of a system is the existence of two or more elements of a system that cooperate via interfaces. It is through these interfaces that a system attains its superiority over an unorganized collection of things. We must conclude that a system must have some interfaces. The richer the interface complexity the greater the potential for synergism between the components.

We know from Chapter 2 that we must decompose a system functionality into system elements that can be designed by specialists. We commonly organize our engineering departments according to how we apply these specialized engineers to designing elements in our company's product line. It is a fact that the majority of system problems will occur at the planes in a system where different specialized organizations are responsible for the opposite terminals of an interface. The way to minimize these problems is to minimize the number of such interfaces. It also helps some to disconnect the functional organization from the product organization by organizing the project by product-oriented teams we have called PDT.

We must have interfaces between elements designed by different agents yet we desire to minimize these interfaces. This is perhaps the most important task of a system engineering organization or the PIT, to control the development of the interfaces in a system to satisfy these conflicting demands. Obviously, the solution will be a compromise between two extremes, a rich interface in a system that will drive you to the poor house to develop and an uncoordinated collection of independent things that does not form a system.

12.4 The solution

The principal technique for arriving at a reasonable solution to the interface dilemma is careful development of the architecture to respect the known design team organizational boundaries. We can allocate the functionality and aggregate the architecture elements so as to minimize the need for cross-organizational interfaces.

As the architecture is created we must continue to analyze its expanding interface implications within the system itself and between the elements of the design team in order to assure that we remain in a condition of optimum interface definition. An effective tool for this purpose is the N-Square diagram. An N- Square diagram is a square matrix with size "N" (N cells on a side). The corresponding units of each axis of the matrix are associated with the same set of architecture elements aggregated into families that are under the same organizational responsibility (commonly these are called subsystems).

The diagonal corresponds to internal interfaces for each item of architecture. The other squares correspond to interface opportunities between the elements. Figure 12-1 illustrates such a diagram. The squares are marked with an "X" where there is an interface between the elements identified on the two axes. You will note that there are two squares, one above and one below the diagonal, for each pair of elements. We choose the cells above the diagonal for interfaces that have their source on left face of a diagonal square to the top of a lower diagonal square and those below the diagonal for those that have their source on the right side of a diagonal square to the bottom of a higher diagonal square. There is nothing wrong with making the opposite selection. Just be sure that you use the same convention throughout the analysis.

As we can see in Figure 12-1, there is an interface requirement between the actuators that are a part of the hydraulic system and the DCU (on-board Digital Computer Unit) that develops the command signal. There are also interfaces between the actuator and the hydraulic control unit that provide hydraulic power to the actuator in response to the control exercised by the DCU. We can use this diagram to evaluate the functional allocation choices we have made against the organizational responsibilities.

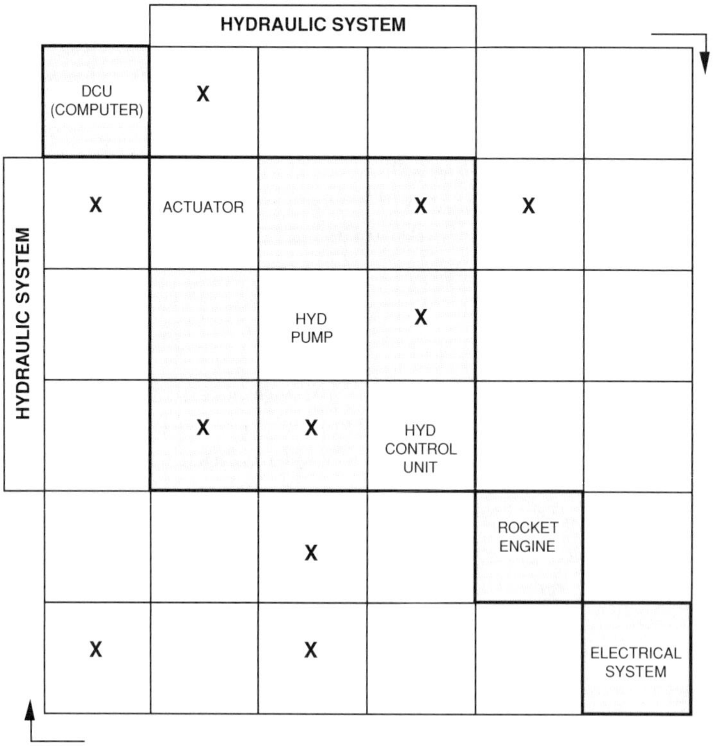

HYDRAULIC SYSTEM

DCU (COMPUTER)	X				
X	ACTUATOR		X	X	
		HYD PUMP	X		
	X	X	HYD CONTROL UNIT		
		X		ROCKET ENGINE	
X		X			ELECTRICAL SYSTEM

HYDRAULIC SYSTEM

Figure 12-1 Typical N-Square diagram.

Let us assume that each of the subsystems we have identified on the diagram is the responsibility of a separate design department or PDT. Therefore, the shaded squares on the diagonal represent interface situations between system elements within a design group responsibility. There is only one design agent responsible for both terminals of these interfaces and they will tend to be properly developed with little outside integration by the PIT or system engineering community. Any interfaces identified outside of these shaded blocks represents a cross-organizational interface of the kind we wish to minimize and the kind that makes our system a system.

As one means of reducing cross-organizational interface, we can reassign an architecture element from one subsystem to another based on the off-diagonal interface count. This is done on our N- Square diagram by moving a row and the corresponding column to a new location in the matrix such that the component in that row and column are now located within a different subsystem. The interfaces will automatically switch to the new configuration. In this case we have not changed the functionality allocated to the system elements, we have reordered the elements with respect to the design team organization.

We may also change an existing interface condition by re-visiting the functional allocations. For example, we could conclude in the case of the actuator that we should have allocated the steering function to a separate steering propulsion system that does not require actuators, only exhaust jets that are turned on and off under the control of the on-board computer. In the process we would eliminate the hydraulic system if it is not required for any other function.

Alternatively, we could change the design concept from a hydraulic actuator solution to an electrically driven actuator solution. Now the complete control system is in the avionics design domain, the computer and the actuator. We may have removed the need for a hydraulic system as well.

These three approaches are effective in refining the system architecture to simplify system interfaces while assuring that we are satisfying the needed system functionality. We will expand on the N-Square diagramming process in a moment and describe a companion technique called schematic block diagramming. But first, let us explore a troubling but interesting question about our organizational priorities.

12.5 A variation on the solution

We may ask ourselves if it would not be possible to forget about how our engineering team is organized at the beginning of the project and organize only after we have created the architecture and understand the interfaces. That is, let the product design our development organizational structure. This way we could form our design team organization to minimize cross-organizational interfaces while focusing our system decomposition process on needed system functionality without worrying about the organizational constraint. This is a case of turning the conventional world upside down, but it is an alternative that we should consider. This approach is the basis for concurrent development or integrated product development as expressed in this book.

Another way to say this is to form integrated design teams composed of design disciplines that match the technology associated with the functionality of the element. Many system elements in complex systems require an appeal to electrical and mechanical engineering anyway, so why force yourself to live with a constraining functional organization? The PDT notion linked to the concurrent development concept are based on exactly this concept of organizing the development team to correspond to the product organization. In years past we got into the habit of functionally organizing ourselves in terms of our specialized knowledge bases. We made it worse by co-locating by functions and mutually withdrawing from the cross-organizational interfaces.

12.6 N-Square diagrams and schematic block diagrams

We will describe two diagrammatic tools for the development of a clear definition of the interface relationships needed between the system elements defined on the architecture block diagram. These tools may be used together with N-Square diagrams used as an analytical tool and schematic block diagramming used to publish the results. Either tool can also be used to accomplish the complete interface definition task, the analytical and the exposition portions. We will also find that an interface dictionary implemented with a computer database can be very useful in controlling the development of interfaces throughout the system.

Our methods will focus on producing some combination of three products:

a. A Schematic Block Diagram (SBD), that defines system interfaces. It uses simple blocks and interconnecting lines to define the existence of interface elements and their terminal relationships.
b. An Interface Dictionary that is a tabular listing of system interface elements illustrated on the SBD (one tabular line item for each line on the SBD) containing the element name, a description of the element, its two terminals and media identification from the Architecture Dictionary, and other reference information too extensive to place on the face of the SBD.
c. An N-Square diagram that identifies interface relationships between a set of system elements in a way that forces us to consciously consider every possibility.

12.6.1 N-Square diagramming methods

One tool that has been found effective in exposing interface requirements between the many elements of a system is a matrix called an N-Square Diagram described earlier and illustrated in Figure 12-2 in a compound form.

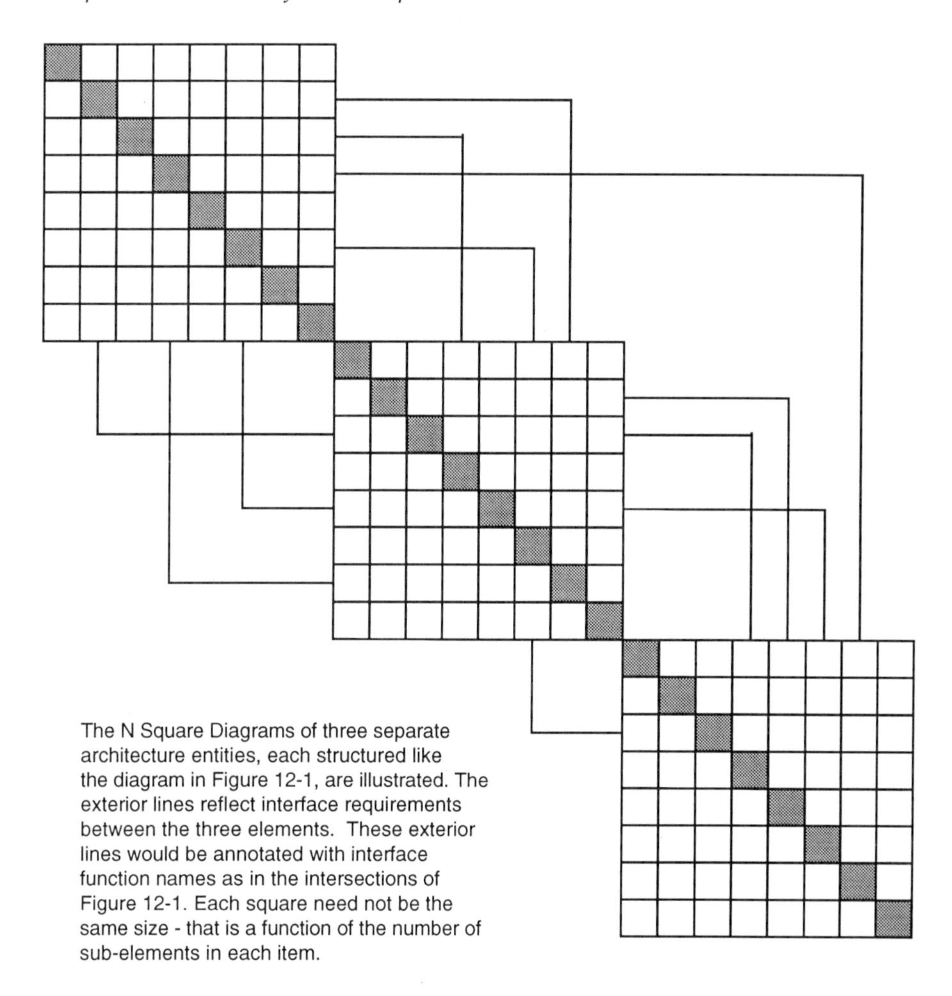

Figure 12-2 Compound N-Square diagram.

The "N" refers to the dimension of the square. To make an N-Square Diagram the analyst makes a square marked off on each side with a space for each system element under consideration. These spaces are then annotated with the architecture ID or element names. The analyst then marks the intersections within the square to note the required interfaces between the elements. The diagonal represents the internal interfaces required for each of the elements and need not be explored in this analysis.

At first glance it is easy to see that there may be a degree of ambiguity in this matrix since it includes two intersections for each pair of architecture elements. This allows us to use the matrix to define interface directivity as well as existence. You have to define which half corresponds to which direction. Establish a convention and stick with it throughout any particular analysis.

The method of noting the existence of an interface in the intersections of the matrix can be as simple as marking the intersection with an "X", using a code with an explanation in accompanying list format, or by writing cryptic notes within the intersections that describe the interfaces needed.

It is possible to arrange a series of N-Square diagrams along a diagonal so as to represent the interfaces not only within particular system elements but between those elements and others as well as shown here. The major elements are broken down into their components on the same diagram. You could conceivably expand this to three or more levels, but the complexity of the diagram will quickly overcome the clarity of its message.

12.6.2 Schematic methods

Two interface illustration methods are employed in the schematic block diagrams (SBD) included here. The first method entails block diagrams where the blocks are the blocks illustrated on the architectural block diagram. These blocks are interconnected with lines that depict the interfaces as explained below.

The second method of interface illustration uses one of two kinds of matrices. When illustrating complex interfaces within a system element a triangular matrix can be used where the same list of elements appear along the vertical and horizontal axes. At each intersection where an interface is required, an "X" is assigned. A dash in an intersection indicates no interface requirement between these elements. The matrix diagonal will always be filled with "X" characters to reflect the innerface of each element listed. Only one half of a square is required because the other half is redundant (directivity not identified).

A square matrix is used to illustrate complex interfaces between two sets of system elements where the interfaces are so numerous that a schematic block diagram would be cluttered by lines so as to loose its utility. In this matrix the elements of two different system elements appear on the vertical and horizontal axes of the matrix. In this case a complete square is obviously required. As in the triangular matrix, a dash in a matrix intersection indicates no requirement for interface between the two elements and an "X" appears in intersections where an interface is required.

Schematic symbols

System SBD illustrations (block configuration rather than matrix) are structured from very simple symbols, blocks and lines. The blocks are the objects, and only the objects, illustrated on the ABD and the lines that join them represent the interfaces between the objects. Figure 12-3 defines all of the symbols used on the SBD.

Interface lines are either of the bundled or elementary type. At the higher levels bundled lines are used denoting the existence of an interface requirement between two system elements. These lines do not include arrows to denote direction because they very likely include elementary interface elements of many kinds with both directions. An elementary line, which

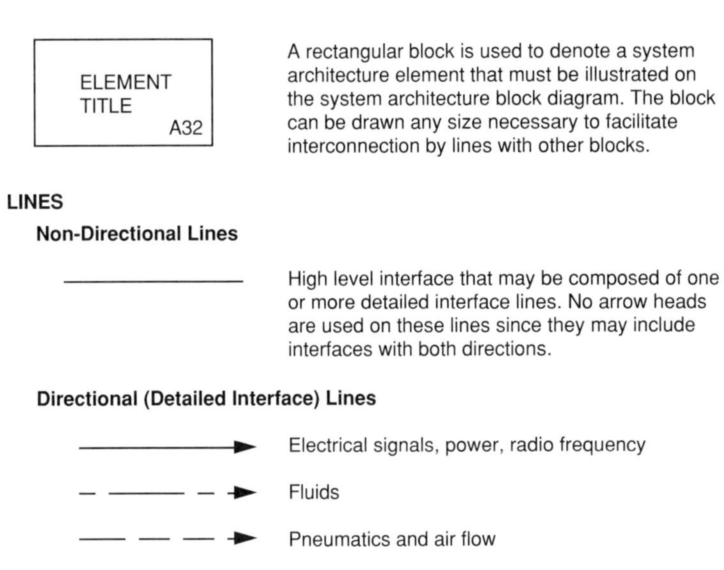

Figure 12-3 Schematic block diagram symbols.

represents a single specific interface, should show direction with an arrow at one end (unidirectional) or both ends (bidirectional).

At the lower tiers where an interface element represents a single elementary interface, interface element names may be included on the line. Also at the lower tiers, different line types may be used to illustrate different classes of interface such as solid lines for electrical, long dashed lines for pneumatic, and alternating dashed lines for fluids, for example. This diagram should be created by the system engineering community or PIT in order to ensure continuity. For example, left to their own devices, the specialized design disciplines will each use the solid line to represent the line they have to use most often because it is easier to draw that any other kind. On electrical schematic block diagrams you will see the solid line for wiring and on hydraulic and pneumatic schematics it will represent plumbing. Each specialized discipline has its own component symbols as well. The system SBD should use simple blocks to represent the system items.

Schematic discipline

It is vital that only those blocks used on the ABD be used on the SBD. If in the functional analysis process, one finds that it is not possible to satisfy system functions currently allocated to system elements as a result of constructing the SBD and the solution is to add a system element, then the analyst must cause the object to be added to the ABD before it is used in the SBD. The interface planes on the SBD must accurately reflect the system architecture and this condition will be realized by respecting this discipline.

Interface coding

Each interface can be coded for unique identification in computer data bases. The author uses the same codes applied to the Architecture ID system. The complete set is identified as "I" which need not be used in cases where the context is clear, but the author retains it for use in computer databases. At the top level the interfaces are arbitrarily numbered I1, I2, I3, ... In. At the next level each of these interfaces are broken down to the next level as I11, I12, ... I1n. This process is continued until the component architecture level is reached in the SBD.

A base 60 system is used here employing the 10 Arabic numerals, 25 capitol English alphabet letters (O excluded), and 25 lower case English alphabet letters (l excluded). The number of characters in an interface ID indicates the interface level. Within each level and branch, up to 60 different interfaces can be coded using this system. If more than 60 are needed they can be broken up into subclasses each no more than 60 in number.

Some people prefer a decimal delimited ID system allowing an unlimited number of identifications within any one level. The code 1.10.4.17 is an example of this method. The problem with this ID system is that it does not sort and index properly in simple computer database systems available for microcomputers. The ID 1.10.4.17 will appear before the ID 1.2.5.7 because a simple database implementation sorts character-by-character and the 1 in 10 of the first ID has a higher priority than the 2 in the second ID.

The work-around often used to make the decimal delimitation system work is to always add spaces to each level corresponding to the maximum number of characters allowed at any one level. For example, we may conclude that there will never be more than 99 elements at any one level. So we would enter the ID 1.10.4.17 into the computer database as _1.10._4.17, where the "_" symbol denotes a spacebar entry. This will sort properly but it is tedious to keep the books straight.

Ultimate SBD

The ultimate, or level zero, SBD is exactly the same for every system and is illustrated in Figure 12-4. Every system interacts with its environment to achieve its top level function (expressed in the system need statement). That environment includes the natural environment and may include hostile systems that the system is intended to destroy or avoid, cooperative systems that provide the system with useful services, and non-cooperative systems that may adversely interact with the system unintentionally (by creating electromagnetic interference, for example).

The system environment is labeled with an Architecture ID of "E" and the system with an Architecture ID of "A". The interface between the environment and the system is labeled here with a level one Interface ID "I2" and the system innerface with a level one Interface ID "I1". In order to reduce the number of levels in the Interface ID, the analyst may choose to assign level one Interface ID to the major interfaces within the top level system elements as well as to the environment interface.

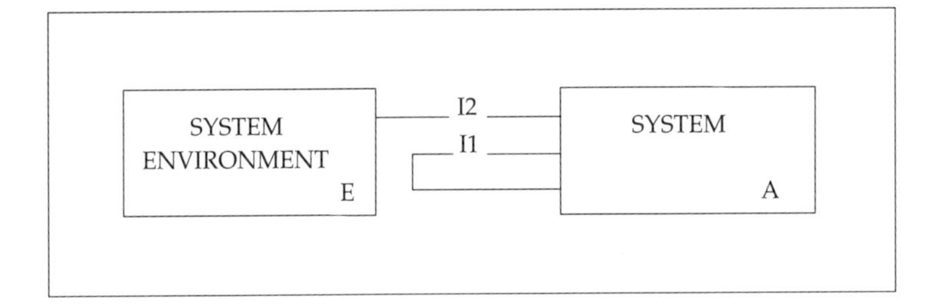

Figure 12-4 Universal ultimate schematic block diagram.

Figure 12-5 illustrates a typical top level (level one) SBD for a space transport system. As you can easily see, this diagram could be fitted into the form of Figure 12-4, which is nothing more than a systems engineering curiosity. The top level SBD is created by laying out each of the top level system ABD elements (as a minimum, those architecture elements immediately subordinate to the system block on the ABD) on a sheet of paper and connecting pairs of them appropriately with interface lines.

Lower tier diagrams must expand the interfaces defined on the top level SBD by creating an expanded SBD for each interface line on the diagram. This process is carried from the top downward progressively in step with the advancing ABD definition. The product of the interface analysis work should be fed back to help define the expanding architecture responsibility assignments.

Note the use of the lines that are attached at both ends to the same block. These interface lines signify the internal interface for that item. These lines correspond to the diagonal on an N- Square diagram. It is useful to illustrate them on a schematic block diagram as a part of the discipline in assigning interface ID codes. The internal interface is coded at the same level as other interfaces at that level. If a schematic block diagram is developed for that item, all of the internal interfaces have the prefix of the internal interface ID. This is the gimmick that assures that interface ID codes can be expanded from the top down in step with the expanding architecture.

Every interface in a system can be said to have a source, a destination, and a media (wires, pipes, attaching bolts, etc.). The intensity of interface problems in a developing system are directly proportional to the percentage of system interfaces that possess development organization responsibility differences among these three interface aspects of source, destination and media. Therefore, one of the principal objectives in laying out the system architecture is to structure the functional elements of the system, driven out by the functional analysis, into subsystems that are cleanly related to the engineering design organization structure.

Interfaces which have different design agencies responsible for source, destination, or media are defined as critical interfaces. These interfaces must be tracked by system engineering to insure they are properly developed by the two or three responsible agencies. Those interfaces which involve different

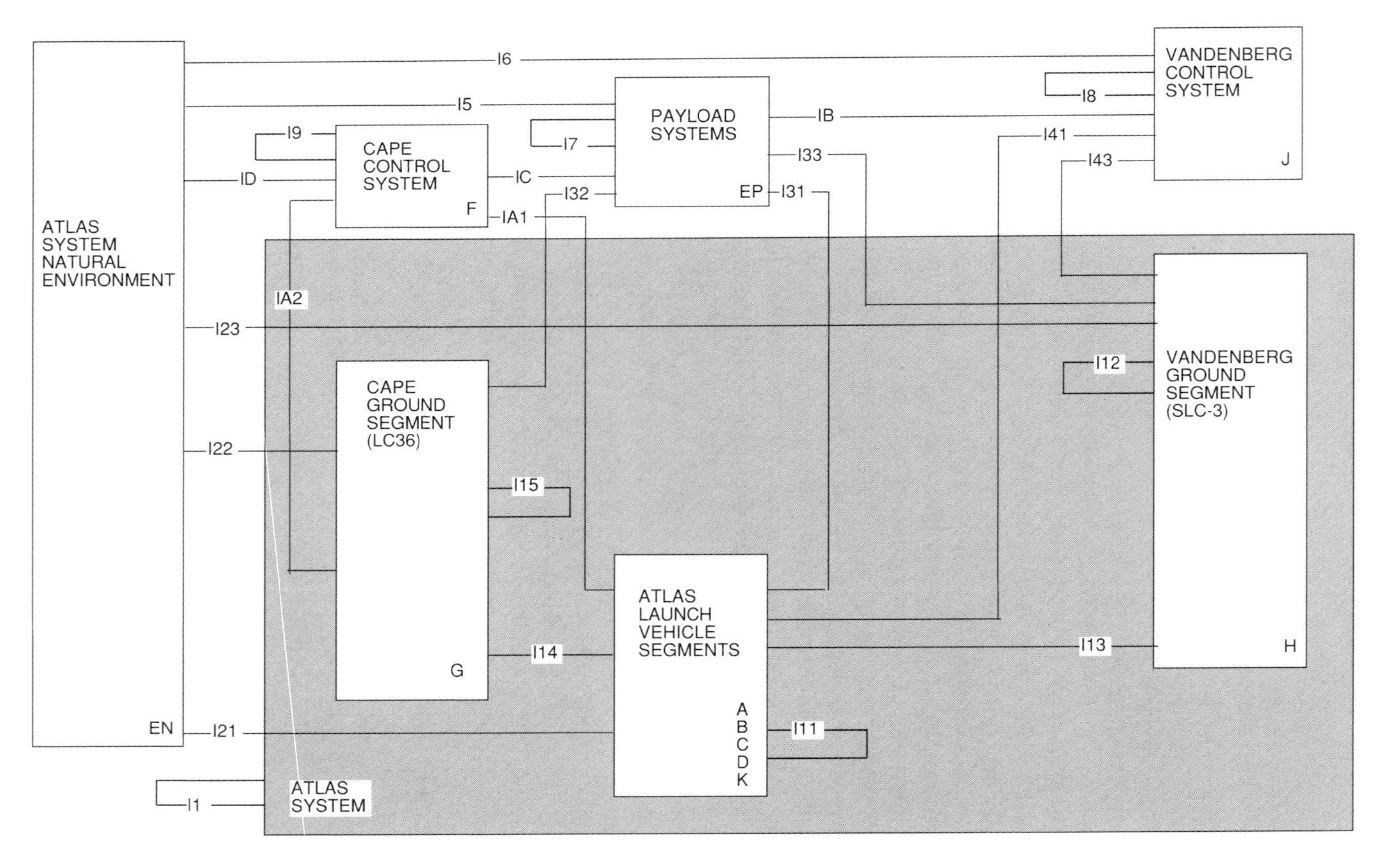

Figure 12-5 Typical system schematic block diagram.

associates or teaming partners should have Interface Control Documents prepared to manage the interface development. Interfaces which have all three aspects (source, destination, and media) under a single design agent's responsibility are non-critical and will be the complete design responsibility of that same design agent.

This objective is satisfied by the concurrent development of the ABD and the SBD under the responsibility of system engineering. As elements are added to the architecture from functional analysis action, they are fitted into the SBD with a watchful eye on alignment between interfaces and design organizational boundaries. This may be done by coloring the SBD blocks different colors for the different design organizations, by overlaying the design responsibilities onto the SBD with dashed boxes, or by simply encircling elements under each design group's responsibility with a line of a particular color (color computer graphics and automated identification of critical interfaces could be a great help here).

The analyst then takes advantage of interface insights provided by the annotated SBD to help guide placement of architecture elements in the ABD so as to result in simple interface relationships between the development organization elements.

It is of crucial importance that the analyst respect the discipline that the SBD use no blocks that are not already identified on the system ABD. The reason for this is that architecture elements can have assigned to them organizational development responsibilities, which, in turn, can be used to define interface development responsibility pairs. If these are not clearly connected, then management of the development process will suffer.

SBD expansion

Later we will refine our definition of system interfaces to identify three subsets: innerface, crossface, and outerface as a function of the principal engineers responsible for architecture development. For the time being let us just say that innerface is those interface elements that have both their terminals at the same architecture element. A crossface interface element has two different architecture terminals. It is useful to have a name for interface elements that are of no interest to a particular development agent, thus the outerface class.

The top level SBD is expanded in step with the expanding ABD. There are two principal expansion approaches: (1) Innerface Expansion, and (2) Crossface Expansion. These two subsets of the complete interface set for a system encompass all system interfaces so it is unnecessary to focus special attention on the third class of interface called outerface. More on this method of classifying interfaces momentarily. Just accept these as names for interface classes for now.

Innerface expansion

Innerface is identified on any SBD as a line with both ends on the same block. This line will be coded like any other with an Interface ID. To expand one of these innerface elements, the analyst should prepare an SBD sheet in either the block diagram or triangular matrix form that illustrates each of the system elements subordinate to the block that has both ends of the line attached. The subordinate blocks are derived from the ABD. The analyst will find that initially it may be useful to simply make a circle on a piece paper for each block. Then draw a line joining each pair that must have an interface. Add an innerface line to each block. Add the Interface ID to each line in an arbitrary fashion by appending 1, 2, 3, through z (using our base 60 system) to the Interface ID for the line expanded in this SBD.

From this preliminary sketch, an example of which is shown in Figure 12-6a, it will be possible to see a pattern of lines and blocks that can be arranged in an artistically pleasing way as well as technically accurate. The final illustration, shown in Figure 12-6b should have as few line crosses as possible. Prepare the final diagram on a computer using a graphics package or using templates and a pencil.

If the rough sketch becomes too cluttered consider using a triangular matrix format to illustrate the SBD. The triangular matrix format corresponding to the interface shown in Figure 12-6 is illustrated in Figure 12-7. It is formed by simply listing the architecture IDs for the set of elements on both axes. For each intersection corresponding to a needed innerface, mark the square with an Interface ID expanded from the original Interface ID.

Crossface expansion

A slightly different technique is necessary for expansion of crossface lines on an SBD. A crossface line will, of course, always have its two ends on different blocks. To expand one of these lines draw a rough sketch with all of the blocks subordinate to each of the two parent blocks illustrated on one side of the sheet. One way to get this sketch is to use the innerface expansions for each of these elements, if they have already been prepared. Next connect pairs of these blocks, one end on a subelement block on one side of the page and the other end on a subelement block in the other side of the page.

When all of the line pairs have been entered, number them all expanding on the Interface ID that is being worked as the root for each of lower tier Interface ID. Use the sketch to see the non-crossing patterns of interest in creating the final drawing. Figure 12-8 illustrates the initial top level, primitive, and final form for a typical diagram expansion.

If a block diagram format becomes too cluttered, use a square matrix format where the blocks of one of the two elements are on one matrix axis and the blocks of the other are on the other matrix axis. Place the Interface ID codes in the appropriate intersections. A dash or a blank space is understood to mean that no interface exists between those two elements.

Compound expansion

There may be cases where the analyst finds it advantageous to combine both forms of expansion discussed above. For example, you may wish to expand a particular interface joining two blocks and the innerface of each block on the same SBD. The advantage to this approach is fully realized only when there are very few other interfaces between those two blocks and other blocks. Avoid preparing schematic block diagram that have so many lines that the value of the diagram is degraded.

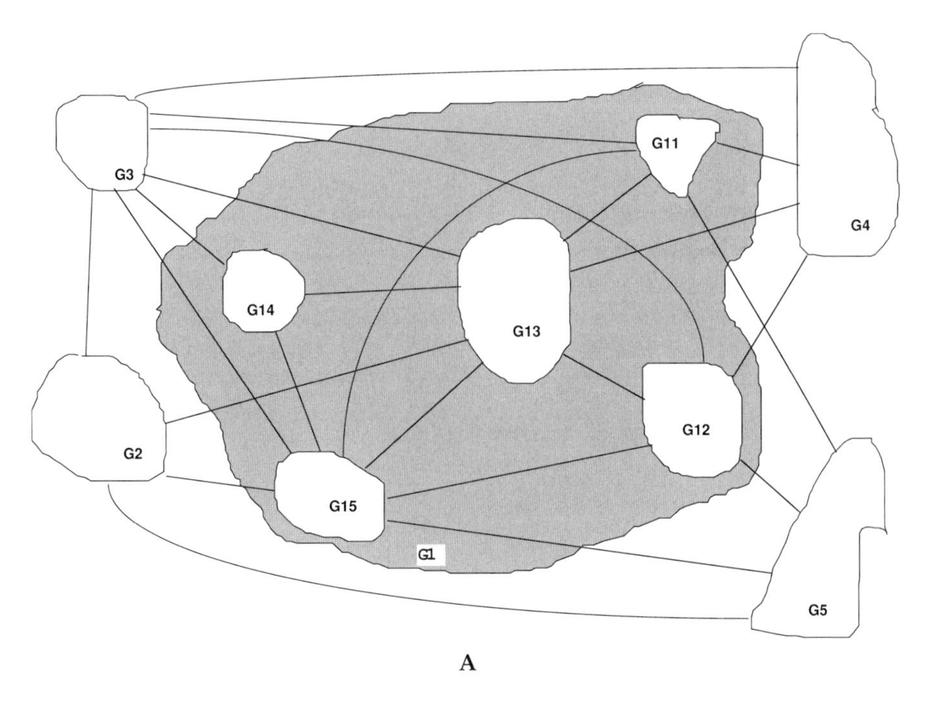

A

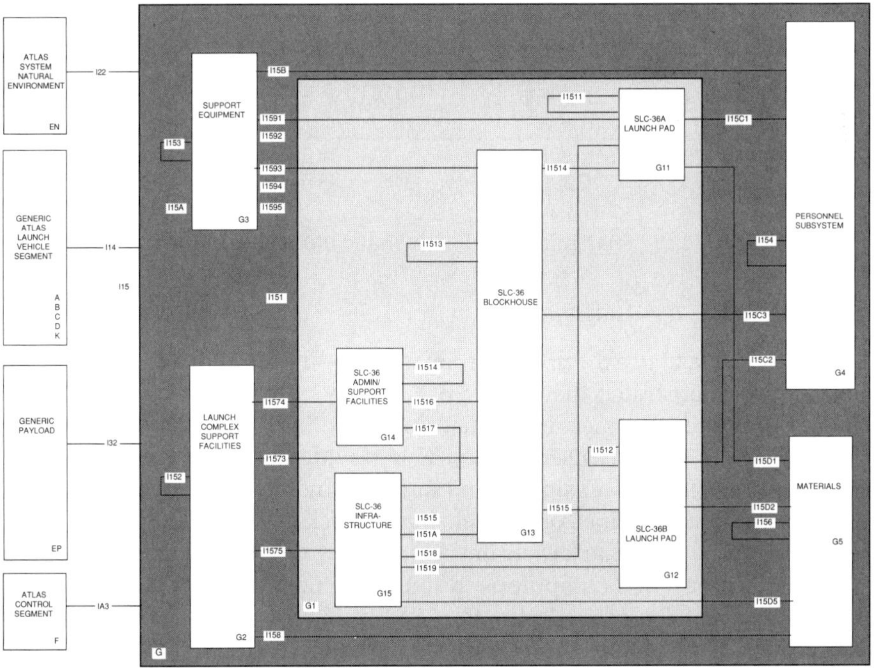

B

Figure 12-6 Innerface schematic block diagram expansion: (A) primitive schematic block diagram; (B) finished schematic block diagram.

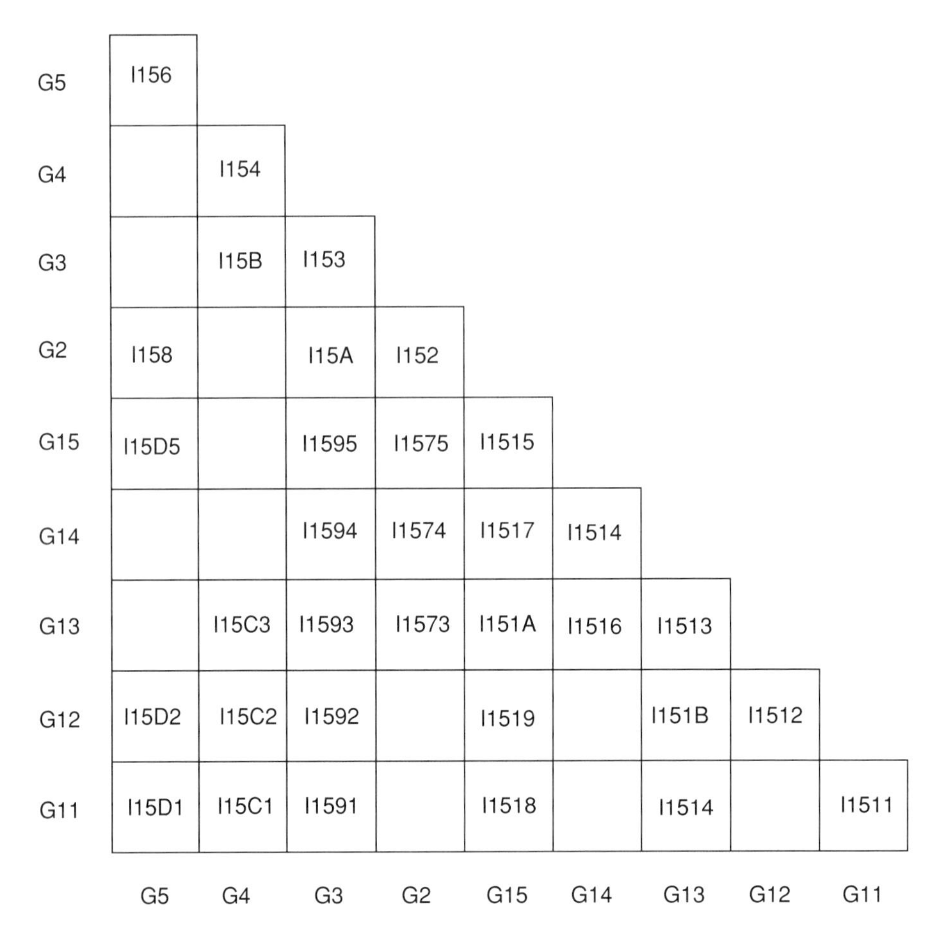

Figure 12-7 Triangular matrix schematic block diagram example.

12.6.3 *Interface dictionary*

It is not possible to use the SBD to retain all of the information of interest about the interfaces illustrated there. The SBD actually only illustrates the existence of interfaces and identifies unique names for each interface. The analyst must also prepare an Interface Dictionary which provides an inventory of all system interface in the form of an alpha-numeric listing by Interface ID corresponding to all of the interface lines on the schematic block diagrams.

Ideally, the dictionary is retained in and printed from a computer database that is a part of a computerized requirements system. It includes information about the source, destination and media of each interface element plus other data of interest. The database allows system engineering to map all interfaces to design responsibility groups for the purpose of identifying the critical interface subset as well as providing interface definition and statusing data. Figure 12-9 illustrates the interface dictionary columns that define each interface in terms of its source, destination, and media. Other columns can be included for other purposes.

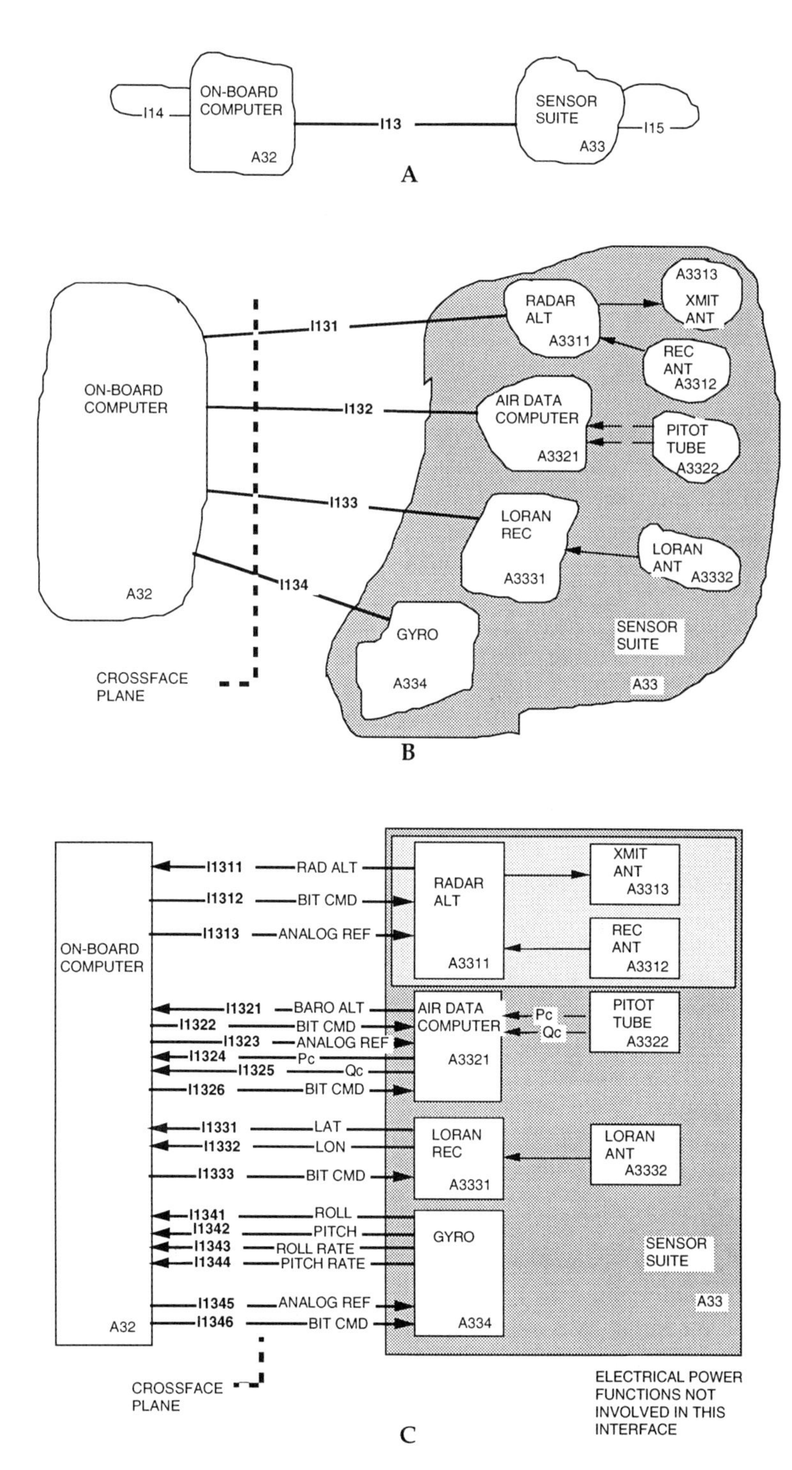

Figure 12-8 Crossface schematic block diagram expansion: (A) top level schematic representation; (B) primitive schematic block diagram; (C) finished schematic block diagram.

INTERFACE IDENT		INTERFACE CONNECTIVITY		
ID	NAME	SOURCE	DEST	MEDIA
I1511	Fuel On Command	A21	A251	A32
I1512	Fuel Flow Light	A251	A21	A32

Figure 12-9 Typical interface dictionary listing.

12.7 *Three views of interface*

Interfaces exist between objects of systems for the purpose of introducing synergism between the objects. A system is said to be a collection of two or more objects that interact in purposeful ways to achieve an objective that no subset of the objects could otherwise achieve and it is through the synergism provided by interfaces between the system objects that this takes place. Interfaces communicate information from one object to another which the other would have no way of obtaining through its own unaided capacity. They provide energy with which to function as in the case of electrical or hydraulic power. Interfaces act to physically tie a system together where objects are bolted or otherwise joined together.

An interface element has two fundamental characteristics: (1) a terminal at each end and (2) a media of communication. The terminal will always be an element of the system depicted on the system Architectural Block Diagram or an element of the system environment. This includes the elements of the system of interest, possibly some cooperative and hostile systems within the system environment, natural environmental elements, and no other possibilities. Likewise, the media of the interface element will be provided by some element of the system of interest, or the system environment (natural environment or cooperative systems). Wire harnesses, fluid plumbing and the space through which radio signals pass are examples of interface media. The first two of these media are provided by the system architecture and the last is an example of an element completed by the system environment.

We take three different views of the system interface here as a function of the perspective of the observer. These three views take into account the variable intensity of interest and responsibility for interface by the design community. The basis of these differing views is the relationship between the terminals and media of an interface element and the responsible design organization. It is a great systems management truth that the interfaces that will result in the greatest system problems are those where different organizations are responsible for two or three of the following: source terminal, destination terminal, and media. Relatively little difficulty can be expected where the same organization is responsible for all three. Where differences occur, the interface is a member of the cross-organizational interface set that system engineering must focus its energy upon in interface integration work.

The traditional view of interface is that an interface exists where two different contractor organizations are responsible for the design of the two terminals of the interface. We take a more general view here where an interface is any means of relating one system element to another at any level of system indenture. Interface thus imagined may be partitioned into subsets of interest as a function of organizational responsibility by establishing three classes as follows:

a. OUTERFACE — Interface elements with neither terminal or media under the design responsibility of organizational element X are the organization X outerface class. Organization X has no immediate interest in this class of interface and is not responsible in any way for its proper design.
b. INNERFACE — Interface elements with both terminals and the media under the design responsibility of organization X are in the organization X Innerface class. Only organization X is immediately concerned with this class and is completely responsible for its proper design.
c. CROSSFACE — Interface elements with some subset (other than the null set) of two terminals and media under the design responsibility of organization X and the remaining subset (not the null set) under the design responsibility of a different organization are in the organization X Crossface class. Organization X and some other organization are jointly responsible for the design of this class. This is the interface class where system engineering energy and budget must be focused because it is where system problems will develop. These are the cracks between the specialized knowledge and experience of the engineering organizational elements. They are inevitable in any system and are determined by how the organization is structured with respect to the structuring of the system architecture. It is important that the system engineering community be capable of sorting all of this class of interface from the whole and assigning responsibility for its development.

12.8 Interface responsibility model

How shall we assign responsibility for interface development? It is clear there is a fundamental difference between architecture development responsibility and interface development responsibility. The former can be completely assigned to a single development agent with little difficulty. Interface on the other hand has two, and with the interface media possibly three, agents that should be interested in any one interface element. We need a way to make interface responsibility unambiguous.

The previous discussion of the three classes of interface as a function of your outlook offer a solution. First let us create a hypothetical schematic block diagram in Figure 12-10. It includes 7 components in three subsystems. Our challenge now is to determine who, among the 7 principal engineers at the component level and the three principal engineers at the subsystem level, is responsible for the each of the interfaces illustrated here.

The system illustrated in Figure 12-10 is composed of a set of architecture consisting of A={A11, A12, A21, A22, A31, A32, A33} arranged into three

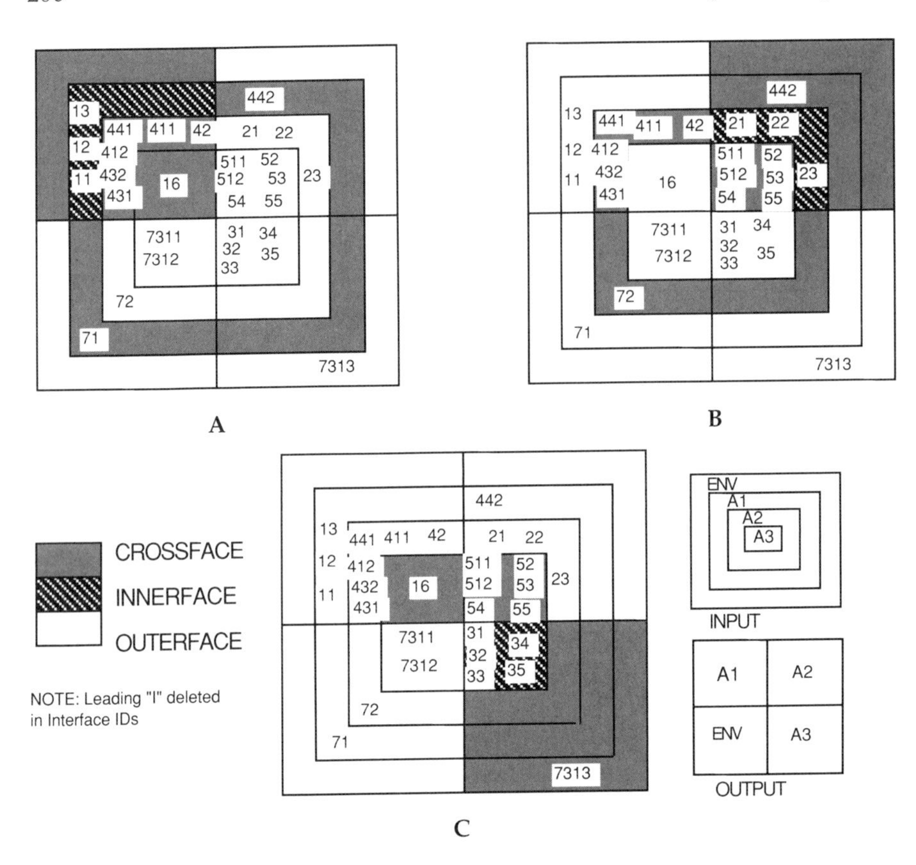

Figure 12-12 Subsystem principal engineer views: (A) subsystem A1 perspective; (B) subsystem A2 perspective; (C) subsystem A3 perspective.

12.9 The special need for external interface development

Interfaces between elements that are both under the same company's design responsibility should never be a problem in a development activity, but they commonly are even where different design departments of the same company are involved. Interfaces between elements a contractor is developing in-house and those that are procured from an outside vendor should never be a problem since the contractor controls the vendor through a subcontract or purchase order. Interface definition is a common area of dispute between contractors and their suppliers, however. If interfaces that are fully under the control of a contractor are difficult to develop flawlessly, those between two contractors which have no contractual relationship could be nearly impossible. Because it is so difficult to develop these interfaces, we need some special controls at these very important interface planes.

Interface Development is an engineering management and technical approach to the development of interfaces where the contractors responsible for the design of the elements at the two terminals are different and not linked through a contractual arrangement. Commonly these contractors are called associates, and each has an independent contract with a common contracting agency. This management approach seeks to introduce discipline and precision into the technical communications between the contractors and to formally resolve problems across the interface between their products.

A customer will commonly require two interfacing associate contractors to reach an agreement on how they will jointly work a mutual interface and to jointly sign a memorandum of agreement stating the means by which they will cooperate. They may also be required to develop an interface management plan to expand on the memorandum of agreement and both sign that plan. One of the associates will be required to prepare an Interface Control Document (ICD) and they and the common customer will manage the development of the interface through that ICD.

An Interface Control Working Group (ICWG) will be formed with the customer chairing its meetings. The ICWG will meet periodically and take up interface issues. Over time, the interface definition evolves between the two elements managed in this way. Interface issues will be resolved by joint action and approved as evidenced by approving signatures on the ICD and revisions (Interface Change Notices or ICN) by ICWG members.

Two possibilities exist for this ICD. One possibility is that it will only be used as a means to an end, that being to provide each associate with the same interface definition for use in their design work. In this case, each associate uses the ICD as the source of the interface requirements for the specification on the element on their side of the interface plan. When these two specifications are authenticated (signed) by the common customer into the program baseline, the interface can thereafter be managed though engineering change proposals (ECP) against the two specifications. The ICD goes away at this point having served its purpose.

The other alternative is that the ICD remains a living document throughout the program. In this case, the interface requirements contained in it may never be copied over into the specifications at the terminals. Each terminal specification will reference the ICD for the interface requirements.

These is a third alternative that is nothing but trouble and that is the case of a living ICD and individual contractor attempts to capture the interface requirements in their terminal specifications. This results in a triple set of books that will almost certainly get out of synchronism.

chapter thirteen

Requirements integration

13.1 What is requirements integration?

Requirements are attributes for things (in the system architecture) and processes in the system that must be respected by designers when synthesizing those requirements into a design solution. There are integration measures that should be applied to the process of developing these requirements for individual items or processes and across the whole system composition. Some of this work requires active interaction during the requirements definition period while other work involves auditing the results of requirements work previously accomplished. So, there is a temporal or program phase-related aspect to requirements integration work.

How we assign responsibility for requirements integration work has a lot to do with how we are organized on programs and the range of requirements analysis strategies and methods allowed or used. We have assumed throughout this book that we were organized within a matrix management structure with personnel supervised on programs within PDT defined by program management. The principal discussion in this chapter will focus on this structure.

Requirements integration entails ensuring that: each requirement for an item satisfies a set of minimum criterion, that the set of requirements for an item satisfy a minimum criteria, and that all of the requirements for an item are traceable to other requirements and related information. It is largely a requirements quality control process. The big question is, who is going to be responsible for this work. Some of it should be the responsibility of those writing requirements while other integration work can only be done at the system level by the PIT.

13.2 Requirements integration responsibility

Requirements will flow into system and team item specifications from the many specialists staffing the responsible teams. The aggregate set of requirements for any one item may not be free of conflicts. It is very important that the PIT make it clear who is responsible for requirements integration within the PIT and within the PDTs. A principal engineer should be assigned for each specification and that principal engineer should be made responsible for development of the document content including proactive identification and integration of conflicts.

In addition to performing requirements integration on system level specifications, the PIT should also monitor the requirements integration work done by teams. This commonly cannot be done in total because of budget constraints so any audits must focus on the most critical items and requirements categories.

13.3 System level SRA overview

Figure 13-1 offers a view of the generic PIT system requirements analysis process, during the phases when development requirements for Part I or Type B specifications are being identified. In paragraph 7.3.4 we discussed SRA at the item level within a PDT. The process illustrated here fits into each of the several PDT processes on-going simultaneously via the concurrent engineering bond indicated. Within the PDT the SRA work entails cross-function, co-product integration work while between the PDT and PIT teams the integration work is more focused on co-function, cross-product integration. All of the engineers and analysts involved in the item and system requirements analysis processes should, as we have discussed earlier, have a means of cooperating easily via physical collocation or excellent communications and computer networking of tools and information products.

Refer to the companion *System Requirements Analysis* by the author for a detailed discussion of the process illustrated in Figure 13-1. We are most interested here in the tasks identified as Integrate Product Requirements, Verification Integration, and Requirements Audit and Traceability Analysis.

13.4 Requirements integration activities

Those responsible for requirements integration should perform 5 specific audit actions on requirements described in subordinate paragraphs. You will likely not have the resources to complete all of these audits on all of the items discussed so you must prioritize the work and distribute it such that PDTs are doing the majority of the work and the PIT does spot checks to identify items for further audit action. The content of this paragraph was originally written by the author in the book *System Requirements Analysis* published by McGraw-Hill in 1993. It is included here with the permission of McGraw-Hill.

13.4.1 Individual requirements audit

Each individual requirement should be checked against at least the following seven criteria: traceability, style, singleness of purpose, quantification, verifiability, unambiguity, and good sense. These criteria are from the book *System Requirements Analysis* with some refinement.

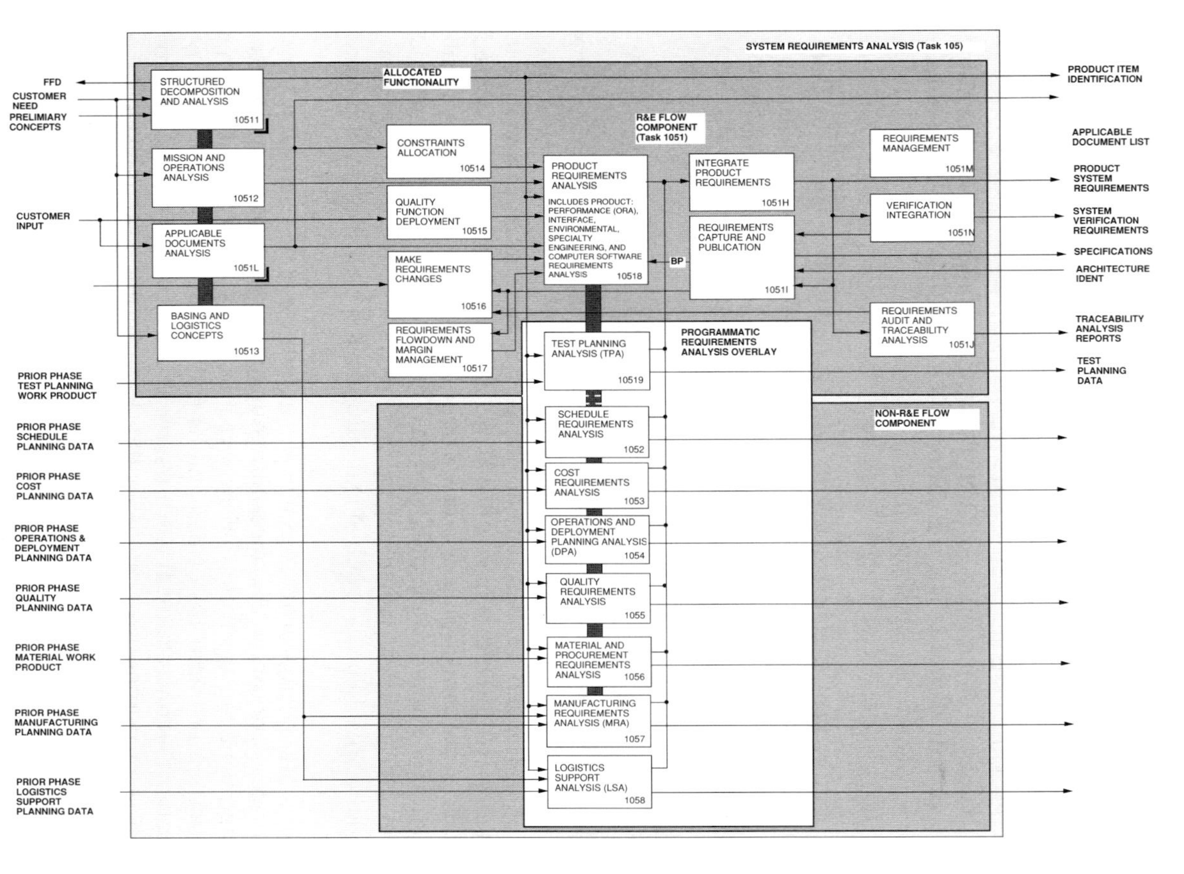

Figure 13-1 PIT SRA Process.

Traceability

Every requirement should be traceable to a driving need in terms of parent requirements, the requirements analysis process through which the need for the requirement was stimulated or a source reference, and the verification process through which it will be proven that the design is compliant. Every requirement should, theoretically, trace up to the customer need. Any requirement that does not upwardly trace is suspect and the rationale for non-traceability should be captured and approved or a condition of traceability established.

Correctness of style

Every requirement should be written with correct language, spelling, and punctuation. It should reflect the customer's style guide and data item description where those are supplied.

Understandability

Every requirement should be easy to understand. Only a single semantic interpretation should be possible. Simple words, simple sentence structure, short sentences, and avoidance of negatives helps to achieve this criterion. Brevity and adherence to the one-paragraph-one-requirement rule also helps to satisfy this criteria, with one possible exception discussed in the next criteria.

Singleness of purpose

There seems to be a tremendous attraction to writing requirements in long paragraphs. It is not unusual to find many requirements in one specification paragraph. The only valid argument for this, and the argument is marginal, is that some requirements have to be stated as two or more closely coupled thoughts in one paragraph and to break these kinds of ideas apart may result in violating the understandability criteria. This should be the exception and will never be a valid argument for paragraphs three pages long. In general, we should strive to include only one requirement in each specification paragraph.

The principal argument for singleness of purpose is that to do otherwise complicates verification and traceability due to the potential for ambiguity. The argument can be made that new requirements analysis tools permit identification of fragments of paragraphs (called parsing) for purposes of verification and traceability. If you are served by such tools, the argument for singleness of purpose may be less valid. Even in this case, however, the author confesses to a bias in favor of simplicity of structure over other considerations.

There is one other problem raised by the singleness of purpose criteria. MIL-STD-490A (and the draft B revision as well), the military standard for specifications, bounds the number of indentures in paragraph numbering to seven levels. Limiting requirements statements to one requirement per paragraph generally will increase the number of levels and commonly beyond the seven level restriction. Adherence to this criteria may require tailoring of the standard on a military contract. Alternatively, we can use lettered subparagraphs to satisfy both the singleness of purpose criteria and the seven level restriction. Once again, if your requirements tool can maintain traceability and verification hooks uniquely to the fragments of interest and generate specifications from this structure, it may be a good overall solution.

Quantification

Requirements should include not only what is required but the needed value or values. Some kinds of requirements must be stated in qualitative terms, but they are exceptions. We should seek to quantify our requirements. Otherwise, it is very difficult to determine if the requirement is satisfied.

Verifiability

One of the most effective ways to ensure good requirements are written is to require that they be verifiable through some practical process. In the process of trying to determine how to verify a requirement, you are forced to write the requirement properly. This is why many people believe that the verification requirements (Section 4 of a specification) should be written by the same person who wrote the Section 3 requirement and that it should be done at the same time. This also encourages that quantified requirements be written because it is very hard to devise a verification process for non-quantified requirements.

Good judgment and good sense

We need to check the requirements set for good sense. Requirements must be written against characteristics of the product and not against things that cannot be controlled with all the money in the world, like the weather. We must not violate the laws of physics or any other discipline. Once again, simple language is a great assist in making this check.

13.4.2 Requirements set attributes

The complete set of requirements for an item should satisfy the following five criteria: consistency, completeness, minimized, uniqueness, and balance. The principal engineer must ensure that these criteria are satisfied during the requirements analysis effort and the management approval process should encourage the principal engineer to give evidence to that effect during reviews. These criteria are from the book *System Requirements Analysis.*

Consistency

The set of requirements is internally consistent if it does not entail self-contradiction. The set is consistent with all other requirements sets if it does not conflict those other sets. An example of inconsistency in an item requirements set would be where one requirement called for non-use of strategic materials and another requirement that cannot be synthesized without the use of such materials.

Completeness

How can we be sure we have identified all of the appropriate requirements for an item? Unfortunately this is a question that cannot easily be answered. But, there is a good chance we will satisfy this criterion if a qualified staff of engineers used a systematic approach to identify the attributes that must be controlled. That is exactly the basis for the structured approach encouraged in this book. If we have conducted a thorough functional analysis of what the item has to do, an effective environmental analysis, a systematic interface analysis, and an integrated specialty engineering requirements analysis and expanded the attributes thus derived into quantified requirements statements using sound quantification methods, there is a good chance that we have satisfied the completeness criterion. In addition to the set including every needed requirement, all TBDs must be replaced with appropriate values and all figures and tables referenced in requirements text supplied and complete.

If we have used freestyle or cloning strategies rather than a structured approach, there is less assurance that we have identified all of the appropriate requirements. A boilerplate can be useful in cross-checking specialty engineering constraint categories of a requirements set generated through the structured approach. Cloning approaches can be effective in specialty engineering and environmental constraints, but are not effective in assuring completeness for performance requirements and interface constraints.

Minimized

But, we just identified one of the criteria as completeness. Now we have a criterion for the opposite situation. Isn't that a contradiction? Strangely enough, it is possible to create too many requirements. Our objective is to decompose the customer's need into a series of smaller problems that will yield to the creative genius of specialized design engineers and their team mates. We need to identify only those requirements that will ensure that the product of the engineer's creativity will work synergistically when integrated into the system.

Requirements have a constraining effect on creativity. We purposely write them for that purpose to ensure that the design solution will have certain important characteristics. Unnecessary requirements constrain unnecessarily and can have the effect of eliminating some potential design solutions that could be better than the remaining options. Requirements do reduce the solution space available to the designer. They should tell the designer the attributes the solution must have in order for the item to function synergistically within the system. They should not define and confine the solution. They should be design-free.

The way to check the need for a requirement is to ask, "What effect would it have if this requirement were deleted? Could the designer, as a result, select a design that would be unacceptable from a system perspective?" If the answer is no, the requirement should be a candidate for deletion.

Uniqueness

Each requirement in the set should be unique in the set with no repetition. Each unique requirement should only appear once in a requirements set.

Balance

Some kinds of requirements are invariably in conflict and we need to find a reasonable balance point in such cases. Reliability and maintainability requirements, as well as cost and most everything else are potential examples. One Atlas rocket had to be destroyed as it diverged from the planned launch path over the Atlantic Ocean. Destruct was commanded by a radio setting off an explosive charge that triggered the propellants causing the vehicle to come apart and its pieces to fall harmlessly into the sea.

There was great interest in what had caused the fault that required the use of the destruct command. Luckily the on-board computer was found washed up on the beach and strangely enough was intact. When tested, it disclosed that a memory location contained data corresponding to the divergence observed and the cause was traced to a lightning strike during ascent from the launch pad. This computer was engulfed in a tremendous explosion, involved in a lightning strike, fell several thousand feet to impact in the sea, and still worked well enough to determine memory content at the time of the incident. The accident investigation team was very happy that this information was available to it, but is it possible that this computer was over-designed? Is it possible that this over-design was driven by requirements values way out of step with real world needs? While it was very helpful to be able to pinpoint the cause in this case, it would not have been unreasonable to expect that the computer would have been totally destroyed.

13.4.3 Margin check

If you use margins to provide slack requirements values to permit easier solution for difficult design problems, the principal engineer should check to ensure that appropriate margin values have been identified, respected, and preserved for the benefit of future program technical risk management purposes.

13.4.4 TPM status check

If there are any technical performance measurement (TPM) parameters selected for the item, the principal engineer should check the history of these parameters and ensure that the goals expressed are feasible.

13.4.5 Specification format check

The requirements must be fitted into a prescribed format defined by the customer for deliverable specifications and an internally defined format for others. As the requirements become available from the analysts, they should be assigned paragraph numbers from whatever format standard is used. If the cloning strategy is used a degree of discipline is automatically encouraged based on the source documents.

13.5 Specialty engineering integration overview

In Chapter 7 we discussed each specialty engineering discipline and how they contribute to the engineering effort. In this chapter we are interested in how the product of these analyses related to requirements definition can be blended into a coherent story. The fact that specialty engineers are tightly focused on their specialty and that frequently the effects of specialty engineering requirements are in conflict, requires that the effects of their requirements inputs be integrated or combined to ensure that a condition of balance is realized in the final design and that an unnecessary cost burden is not placed on the customer. Failure to apply these disciplines to a program can easily produce an unfavorable life cycle cost result due to unsatisfied operability needs, poor match between the system and its environment, and extreme support costs. Application of them with uniformly excessive zeal can result in added non-recurring cost that does not contribute in fair measure to customer benefit. Application of the disciplines with irregular assertiveness can result in unbalanced characteristics of questionable utility. For example, you may end up with an extremely reliable system that cannot, in a reasonable period of time, be tested due to the multiplicity of redundant paths.

One way to integrate the specialty requirements is to hold periodic meetings with the specialty engineers during the requirements development effort and ask each engineer to defend his/her requirements under the critical review of the others. Figure 13-2 offers a more elaborate approach involving integrated checklists. Refer to paragraph 7.1 for the dangers in the checklist solution and paragraph 7.2 for future possibilities

13.6 Interface requirements analysis integration

We need to check that we have identified all of the item interfaces and that we have adequately defined each interface. In addition, we should check the requirements sets for each item interfaced with and verify that their interface requirements are compatible with those for the item in question.

Each principal engineer or integrated product development team should cooperate in the formation of an internal interface working group run by the PIT as a means to evolve a joint agreement on each interface their item has with the rest of the system and the requirements for each interface. With the exception of interface requirements, all of the item requirements are commonly determined within the structure defined by the organization's management hierarchy. We purposely allocate the system's unfolding functionality to items that will associate with how our engineering department or project is organized.

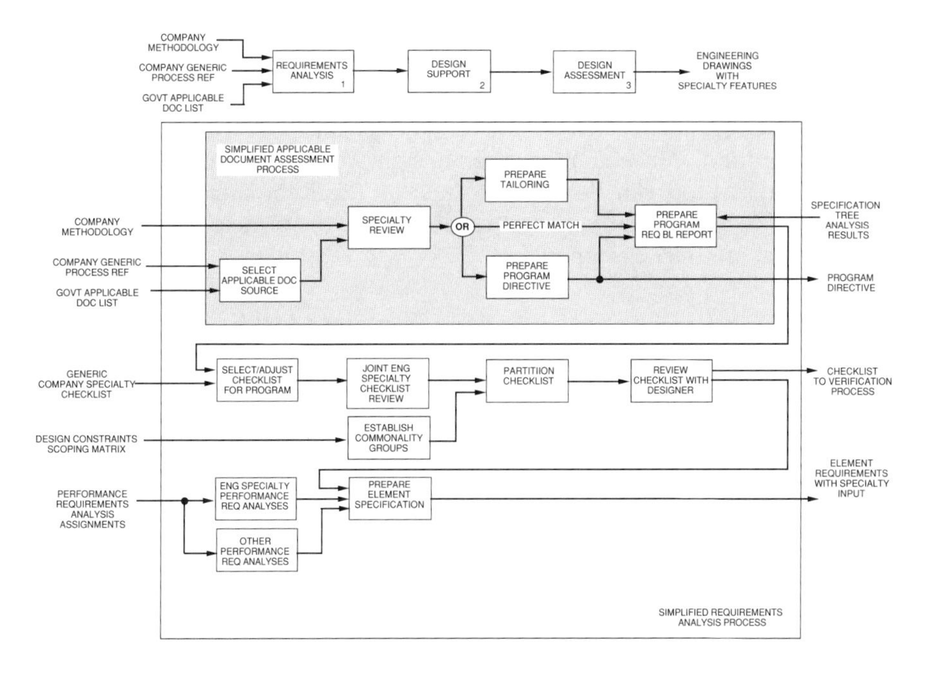

Figure 13-2 Specialty engineering integration process.

Interfaces commonly run at cross purposes with this organizational structure and integration of the interface requirements exposes not only flaws in the product system but in the contractor's program and functional organizations as well. The program team and functional management should be monitoring this process carefully and react to any problems observed to improve the environment for successful teamwork. The symptom to look for is a tendency to withdraw from product interfaces that coincide with organizational interfaces. We need to encourage people to plunge across these gaps from both sides, not withdraw from them. Success here may increase inter-group tension on the program but failure will increase product interface incompatibility problem seriousness. Program organizations focused on the product architecture rather than company functional structure also help to abate this problem as noted in Chapter 12.

13.7 *Environmental requirements analysis integration*

Environmental requirements provide opportunities for much foolishness. Every environmental requirement should be checked to ensure that the language of the statement focuses on product characteristics and not on an attempt to control the natural forces of the World. Also, we should check carefully for conflicts between requirements that are out of synchronization with reasonable expectations (a snow load on the hot tin roof, for example).

In the process of integrating the environmental requirements, one has to determine exactly how they will integrate the union of the different environmental ranges influencing an item in different process steps. A common approach is to take the worst case in all cases. This solution can have a cumulative effect where each parameter is selected this way resulting in over specification and, consequently, over design. We should ask ourselves if some other method can be applied such as the Root Mean Square or a weighted mean. Cumulative worst-case effects is the principal path to components that are over designed.

13.8 Programmatic requirements integration

The same techniques indicated for specialty engineering integration are generally effective for programmatic requirements analysis. The principal engineer must ensure that the whole team remains focused on the same requirements baseline. Frequent informal meetings (that can be brief) of all of the team members that encourage synergism between the members are helpful such that they expose their understandings about this baseline for the education of their team mates and ventilate their own ideas with team mate criticism. Obviously, the principal engineer must encourage a spirit of togetherness and emotional safety on the team or the members will not share their views eagerly and openly.

The team must also have an effective means of communicating the baseline among themselves and with other teams. Common computer databases networked throughout the project, good voice communication capability, and physical collocation are helpful to this end. In addition, a team might use a war room concept to post a lot of information for simultaneous view. It is not necessary to have an actual room for this purpose. If the facility has any large expanses of wall, in corridors for example, they can be used very effectively.

The principal objective in integration of product and process requirements is to identify inconsistencies between the requirements for each. Some obvious case studies of conflict may be helpful in suggesting many other possibilities:

1. The current product requirements can only be satisfied by a new metallurgical technology used successfully to date only in a laboratory environment. The item manufacturing schedule does not include time for acquisition of the special tooling required to perform the related operation and the tooling requirements do not list an appropriate tool needed for the operation. The risk list does not include this potential problem and no technology work is envisioned.
2. The system will require 132 articles of a particular product item and 25 spares at the level of the item in question. The principal engineer finds that material requirements call only for 147 due to an error in addition or failure to update the material requirements from 15 to 25 spares.

3. The wing of the aircraft is 53 feet in span but the logistics people are planning for a hangar with a door opening of 48.3 feet because that was adequate for a previous aircraft concept version which was compatible with an existing hangar structure at 13 sites where the aircraft is to be operated. The logistics requirements have apparently not caught up with the product requirements or have not impacted the product design for compatibility with customer resources.
4. The manufacturing tooling requirements call for a master tooling fitting at missile fuselage station 23.5 as a principal support point during missile fuselage build up. The NASTRAN model does not include this tooling point nor the resultant stresses.

There is a fine line between requirements integration and design integration across the product and process valley. Some of the cases included above are very close to this line and perhaps even across it into design integration. There needs to be this concurrent action between the requirements definition for product and process while the requirements are in a state of flux and this same attitude must carry over into the concurrent design of the product and the process components. In actual practice it may be difficult to separate this into requirements and design integration, but it is not all that important that we do so. It is much more important in practice to have this activity working well for your project or company than to worry ideologically over whether, at any particular point, we are into requirements or design integration activity.

This work requires a green eye shade approach to life and can be done very well by people interested in fine details who incidentally also need to be able to grasp the grand programmatic view. And, yes, there is a shortage of these people in your company. The best source for these people is among your most experienced employees who have not progressed in their careers above their level of competence. Hopefully, you have some of these people on the payroll who have also not been damaged over the years through poor management.

13.9 Structured constraints deconfliction

13.9.1 Can there be too many requirements?

The normal idealized situation with an unprecedented system at the beginning is that the only requirement that exists is the ultimate requirement, the customer's need. We apply the structured decomposition technique to identify performance requirements for things in the system exposed through structured analysis. We identify appropriate constraints as well. The point is that we create the requirements from an almost null condition as a part of the process of defining the needed system.

Is the opposite condition possible at the beginning? Is it possible at the beginning for there to be too many requirements identified such that the solution space is null? It is hard to imagine this condition occurring with an unprecedented system but it could occur with a precedented one.

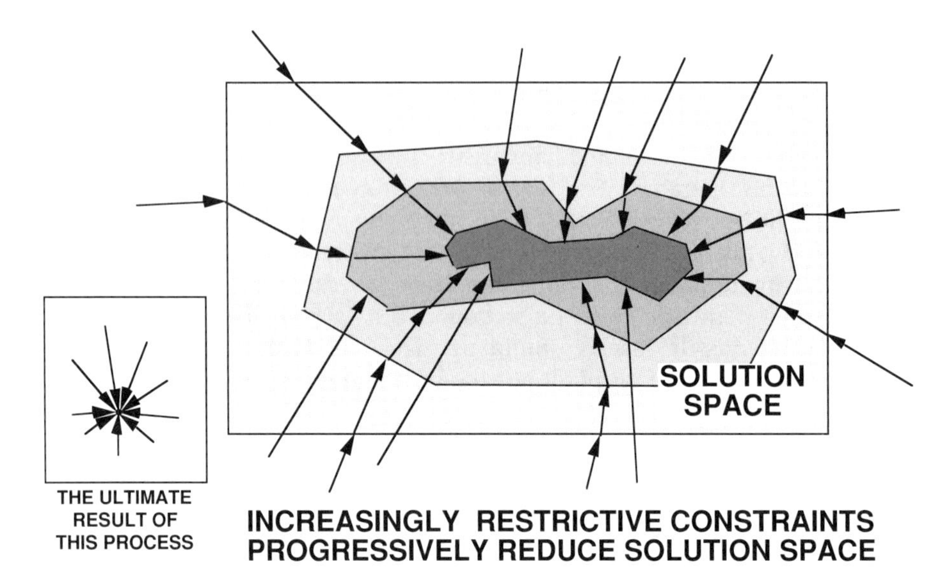

Figure 13-3 Progressively restrictive constraint closure.

An example of such a system is the nuclear waste disposal system in operation at the time this book was published. During World War II when the nuclear weapons development and production process began, everyone was too interested in survival and winning the war to be overly concerned about long term problems resulting from weapons production. This mind set carried on during the selective war with the USSR that lasted until the late 1980s. As a nuclear power industry caught on and created additional waste products, they were stored like those from weapons production.

The end of the Cold War removed the motive encouraging a status quo forcing the Department of Energy (DoE) to place a higher priority on solving the open ended problem of what to do with nuclear waste. Wear-out of existing storage facilities and the increasing volume of the ever expanding material requiring storage also contributed to moving this problem higher on the list.

Here is a system that arrived at a condition of over constraint because of the difficulty and cost of solving the system problem. The constraints applied to the solution space precluded a solution so the problem grew worse over time as shown in Figure 13-3 adding more constraints. An open loop situation like this can only get worse over time, in this example, it has. The ultimate result of this kind of process is a null solution space.

13.9.2 Deconfliction

An over constrained condition can be approached in a structured fashion. First, we identify all of the constraints that apply to the system and link with those constraints to the key stakeholder who insists on that constraint but

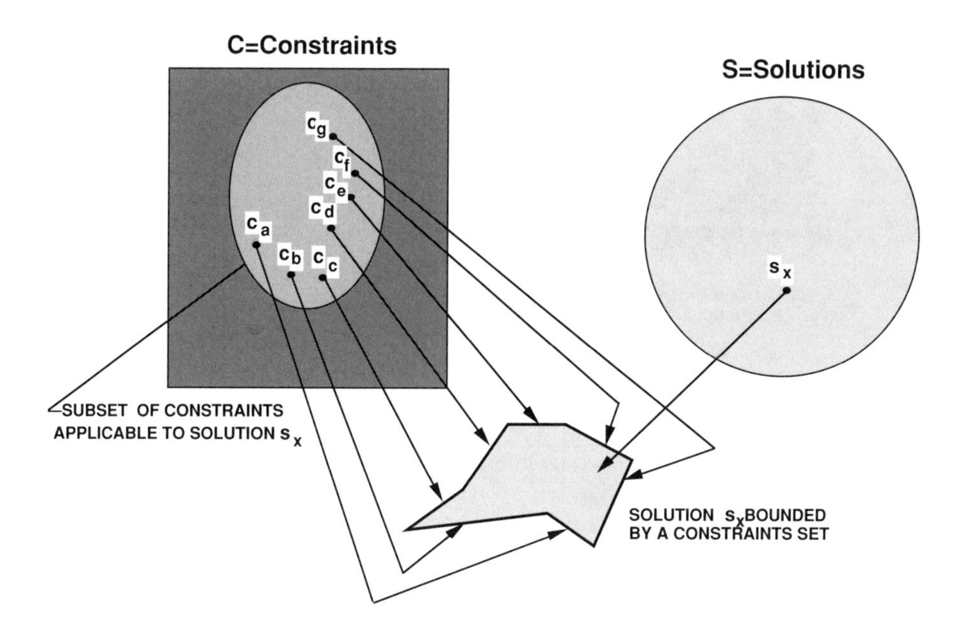

Figure 13-4 Solution-constraint mapping.

also has the power to change it. It is possible that this is two different people or agencies, but for the sake of simplicity assume they are one in the same. We also need to identify one or more alternative solutions to the ultimate requirement each in the form of a scenario, mission statement, or design concept. We can then map all of the constraints to the alternative solutions. We would be drawn to this mode of analysis because all of these currently defined solutions had already been declared unacceptable for none of them meet all of the identified constraints. The result is illustrated in Figure 13-4. Our goal is to expand the solution space for one or more solutions such that a solution is feasible.

At this point we must make some decisions about how much energy and money we wish to pour into the analysis. We can pursue all of the possibilities or exclude some alternatives based on a cursory review. For all surviving alternative solutions we now must evaluate the difficulty of changing each applicable constraint sufficiently to permit that alternative to be successful. We have a simple book keeping situation now with lists of things to do and these can be assigned to teams of people. Figure 13-5 shows one possible list arrangement for which a computer database would be a natural implementation. For each constraint-solution pair we note the stakeholder, source of constraint authority (reference), the degree of difficulty anticipated in relaxing the constraint, and the effects of the relaxation. We would probably wish to capture other information as well to support the study and a final decision.

At any point in our analysis there will always be at least one solution based on particular values of the constraints, where one of the constraints is how much money we do not wish to exceed, but we may not like any of the

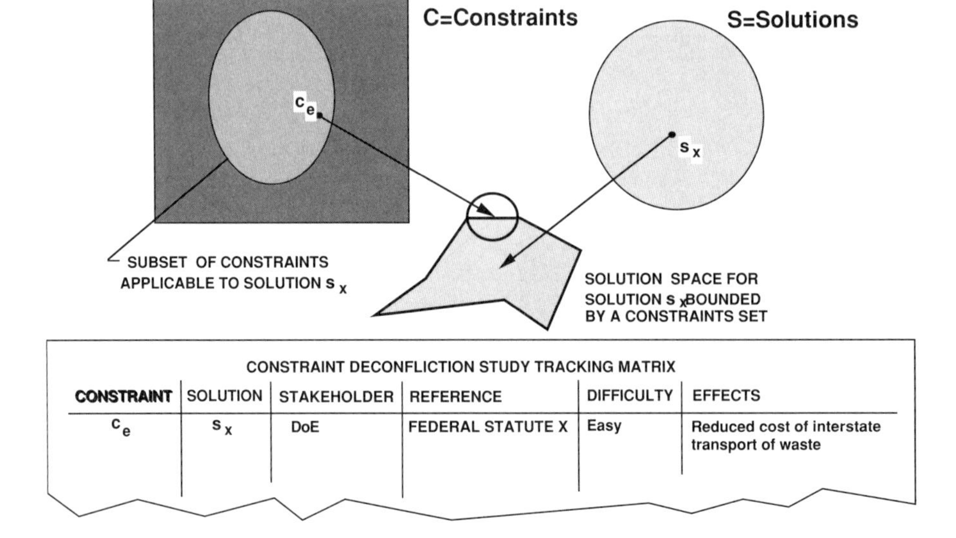

CONSTRAINT	SOLUTION	STAKEHOLDER	REFERENCE	DIFFICULTY	EFFECTS
c_e	s_x	DoE	FEDERAL STATUTE X	Easy	Reduced cost of interstate transport of waste

Figure 13-5 Constraints deconfliction tracking matrix.

possible solutions exposed. At a particular point the remaining difficulties might be that the solution costs too much money, it will take too long to implement, or it exposes people to health risks. What ever the remaining difficulties as we work to roll back the constraints, while the solution space may still be over constrained, it will be changing in the opening direction. From time to time we may establish baseline value combinations for the constraints and then continue work to deconflict the remaining constraints applied to one or more alternative solutions.

Some problems, it is true, could be so difficult that we eventually simply must walk away from them with the conclusion that the solution space is denied because our need is in conflict with the laws of physics, the laws of man, or the aggregate sense of the public good. A problem like nuclear waste disposal, while it is very difficult, we cannot simply walk away from it without placing ourselves in great peril. A problem like this requires an organized search along the lines discussed above where the constraints are consciously evaluated for roll-back possibilities.

chapter fourteen

Product design integration in a PDT environment

14.1 What is the principal problem?

If we have defined the requirements well and planned the development process well and execute the plan faithfully within an environment identical to the one for which the plan was intended, the integration process during design development will be very simple. The reason for this is that there will be no discontinuities, as discussed in Chapter 9, during the design process because we will have perfectly understood the problem prior to beginning the design work. This possibility of future perfection is music to the ears of those who have experienced several imperfect program efforts. But, the realities of our past experiences reduce the credibility of our ever attaining perfection. While development of requirements before design, sound planning, and effective concurrent engineering practices will take us far in our quest for perfection, we should not totally ignore the potential for errors and changed circumstances and fail to provide for course correction as noted in Chapter 9.

Given that we have organized our program by PDTs, coordinated our PDT structure with the system architecture, and provided for supplier and customer involvement in these teams, we will have produced the perfect environment for product integration. We now have to implement, which, unfortunately, is even harder than organizing for integrated development. As we said in Chapters 2, 3, and 12, the principal mechanism of integration is human communications. In Chapter 5 we discussed some specific communications aids that we assume we have taken advantage of.

It is a valid premise, and perhaps an over used cliché, that the biggest problems in design development will appear at the interfaces where different parties are responsible for the development of the design. In Chapter 12 we used the term cross-organizational interface to name this kind of critical interface. We want very much to know where these interfaces are in the system with respect to the PDT responsibilities. In fact, in Chapter 11, we especially defined the evolving system architecture, and aligned our PDT with the product architecture, such that we minimized cross-organizational interface density within the system. You will recall that we cannot eliminate this kind of interface because the richness of the system capability depends on these interfaces as well.

The reason that we worked toward minimized cross-organization interface was to reduce to the maximum practical extent, consistent with maximum system capability with respect to its requirements, the communications problems during design development. Our theory is to minimize the need for the teams to communicate while we maximize their capability to communicate as discussed in Chapter 5. We do this to fight against the potential for the humans to withdraw from the critical interfaces rather than to dash across them to understand their fellows problems.

This is a valid concern. We can overcome it by the techniques discussed above and by clearly defining the interface responsibilities for each PDT. Each team must be held fully accountable for their innerface (both terminals on items within their responsibility). We must allow them to avoid confusion with their outerface (neither terminal on an item under their responsibility), and hold them jointly responsible with another PDT for their crossface. It is the aggregate crossface at the PDT level that constitutes the cross-organizational interface for the system and the principal area of employment for system engineers performing integration work on the PIT and PDTs during product design development.

There should be a specification crafted at the PDT level defining the interfaces with items under the responsibility of other teams, associates, the customer, and major suppliers. These PDT level specifications will define the interfaces in question. Each team must now clearly understand who is responsible for the other terminal of each of these interfaces defined in their specification and accept the responsibility to jointly work with that party for the integrated development of the design solution on each terminal. Methods were discussed in Chapter 12 for bringing this condition about. If we are successful in applying this method to every cross-organizational interface our past product integration problems will be overshadowed by other, less severe problems.

14.2 *How do we accomplish crossface integration?*

Chapter 12 includes some suggestions and tools for accomplishing integration across the cross-organizational interfaces. We absolutely must organize the interface responsibilities and make them perfectly clear. The fundamental key once again, however, is human communication. Every cross-organizational interface element must be the subject of one or more conversations or meetings between the opposing parties with an effort by each party to understand the interface from the perspective of the other's position. The team leaders must also be attentive for signs that team members are unconsciously withdrawing from these interfaces.

Cross-organizational interface is an area where an on-line development information grid (DIG), as described in Chapter 5, can be very helpful. Each

PDT should have a set of interface design charts on the DIG during concept development and preliminary design. Each pair of teams with a joint responsibility for cross-organizational interfaces, as defined by the PIT, must be held accountable for being familiar with their opposing party's interface concept/design and the degree of compatibility of that design with their own. The two teams must also meet from time to time to discuss and resolve interface issues that may be exposed through this joint cross review of the others interface concept.

Each pair of teams responsible for a cross-organizational interface must reach agreement on the existence of a particular set of interfaces between them under the encouragement of the PIT, mutually agree on the requirements for those interfaces, have a means to detect when and if their design efforts come to cross purposes with the agreed upon requirements, and a method by which they may come to a mutual agreement on any actions necessary to resolve any issues across their shared interface.

These are all actions that can be done by a pair of teams. A moderately complex systems may require several of these relationships. Let us say that our program has 5 PDT and that interfaces exist between every pair of items corresponding to our teams. This makes a total of 10 interface relationships that must exist between the teams. Each team will have to maintain 4 interface relationships. Figure 14-1 shows the relationship for two through 8 PDT. As you can see, the maximum number of interface relationships that any one team must maintain is $N - 1$, where N is the number of teams. The maximum number of program team cross-organizational interface relationships is the number of N things taken two at a time or $N!/2(N - 2)!$ (where, for example, four factorial, 4!, is $1 \times 2 \times 3 \times 4 = 24$).

As the number of teams increases, obviously the difficulty of performing product design integration increases. But, the problem is not as bad as the mathematics predicts. The reality is that many of the cross-organizational interfaces will be voids, that is, no interface requirement, as indicated by a

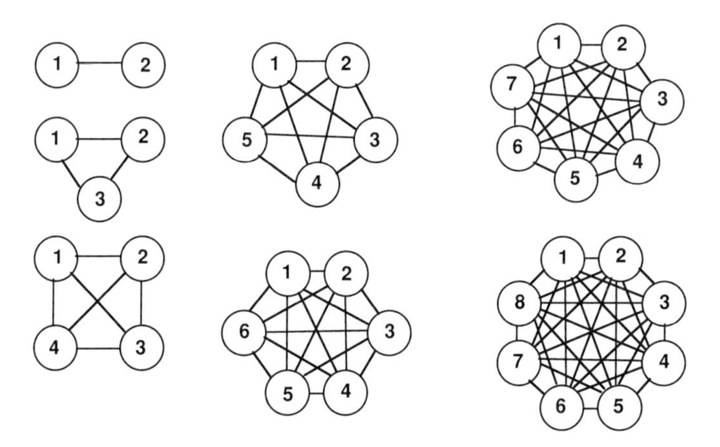

Figure 14-1 Interface possibilities explosion.

void in the corresponding N-square diagram cell (or the absence of any lines between two items on a schematic block diagram). By the way, you should note that a 5-square matrix has 20 non-diagonal cells which is twice the number predicted by the formula above and in Figure 14-1. You will recall, of course, that the N-square matrix includes two cells for each pair to account for directionality. So, all of these techniques for identifying the number of possible interfaces are consistent.

As the number of teams expands beyond some number, perhaps two or three, there must be an outside agency, the PIT in our story, to manage the interface development activities, to audit team compliance with their interface responsibilities, and to track and resolve difficult interface problems. These cross-organizational interfaces must be singled out by the PIT for auditing with the PDT responsible. The PIT must assure that the integrating discussions and meetings are taking place between the teams and the PDT responsible have reached, or will reach, a condition of compatibility.

14.3 There are more interfaces

The PIT must take full responsibility for interfaces external to the system for which the program is responsible. These interfaces may be with associate items, hostile systems, non-cooperative systems, or the natural environment. Where associates are involved, the interface development process may require an interface control working group (ICWG) to formally resolve the interface design. These are all cross-organizational interfaces at a higher level than those discussed in the previous paragraph.

Each PDT should replicate the cross-organizational interface solution discussed above at their own level. Where one PDT has two or more subordinate PDT reporting, they should hold each subordinate team responsible for their component of the major teams responsibilities. This pattern could be broken down further but at some point we must come to the major system components like computers and engines which may have a principal engineer named. This principal engineer, likewise must be held accountable for all of his/her innerface, protected from any responsibility for the item's outerface, and held jointly responsible for developing the crossface at that level. The PDT within which that component fits must act as the PIT for everything within the team's area of responsibility.

14.4 System optimization

The handmaiden of system decomposition and interface integration is system optimization. We carve up the system functionality into an architecture assured that the aggregate of the elements will satisfy the system need in an optimum fashion entailing maximum capability for minimum cost. We may not have allocated the needed functionality so as to yield a well optimized system. It is also possible, during the design evolution, that one or more PDT may develop a design solution, within the boundaries established by their requirements, that places the overall system in a condition of non-optimization. The design solution in one team may force a solution on another team that is unnecessarily costly or force a severe weight penalty on another team. Each team should be forced to carry a reasonable share of the design difficulty if possible.

It is relatively easy to detect the symptoms of non-optimization during the design process. One or more teams will be struggling to meet their requirements while one or more others are having no trouble satisfying their milestones, possibly under-running team budget and schedule allowance in the process. Mathematical models developed to solve requirements and design problems are also excellent tools to help identify sub-optimization as well as to look for optimizing alternatives within the bounds for which the models are appropriate. The PIT must be alert to detect these symptoms and react to them, to understand the driving forces and bring about a balanced condition between the teams if possible.

It may be possible to re-balance the requirements for reliability, mass properties, design to cost, or other parameters such that the teams are equally sharing the burden and a more balanced system will result. If this cannot be done, it may be necessary for one or more teams to alter their design concept or even the technologies to which they are appealing. These actions can ripple into the risk assessment and reshuffle those priorities.

On the other hand, it may not be possible to re balance the pain across the teams, but it may be possible to identify areas of growth potential, corresponding to the over-achieving team areas, that the customer may choose to exploit while providing additional funding. It is also possible to reduce the budget and schedule allowance for one or more over-achieving teams leading to increased margin (or profit) and schedule slack that may have to be spent at a later date to cure a discontinuity. The point is that we should be actively seeking out these conditions rather than letting the corresponding opportunities slip through our fingers from indifference. The alternative to an active program of optimization is that sub optimized conditions will materialize in a time of panic and have to be solved by throwing money at the corresponding problem.

14.5 *Other PIT actions during design*

In addition to the top level interface integration and optimization work for which the PIT must accept full responsibility (task 14111 of Figure 14-2), the PIT must also provide all of the integration services noted in Figure 14-2. These include test planning integration (tasks 14113 and 14114), system level analyses (task 14112), the risk management program (task 14113) discussed in Chapter 9, and the several non-engineering integration functions noted.

The PIT must be constantly alert for opportunities to integrate the design work between product and process. The production process member(s) of

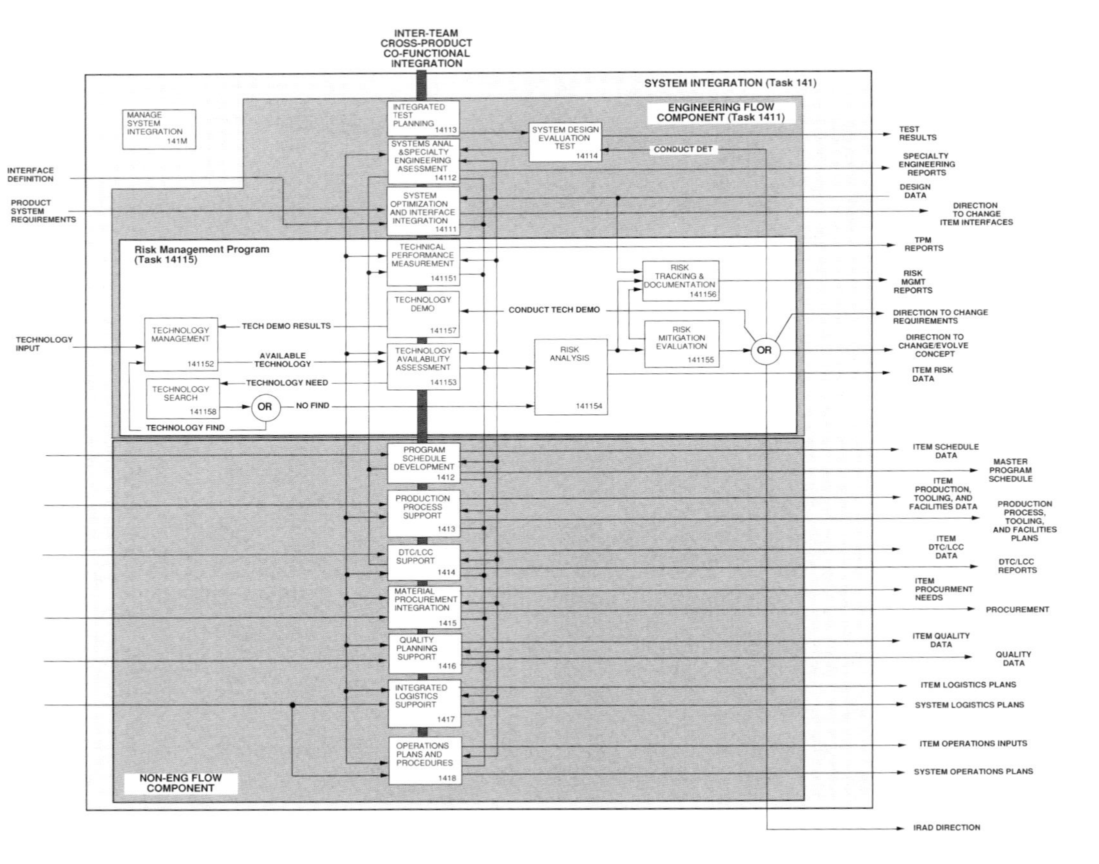

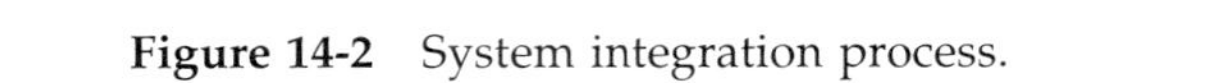
Figure 14-2 System integration process.

the PIT must remain aware of the unfolding production solutions from the PDT and integrate them into a system production solution. At the same time, they must remain alert to possible opportunities to better balance the design solution between the product, operational employment, and production.

This same attitude must be present in the material procurement members, quality members, integrated logistics members, and others. Each team member must remain attuned to what their counterparts are doing on all PDT and integrating those solutions into the system solution while working as a team at the system level to constantly question whether they have achieved the best condition of balance between the evolving designs for the product, the production process, and operational and logistics deployment considerations.

14.6 Special hardware-software integration needs

All of the previous content of this chapter applies to computer software as well as hardware, but software does impose some special concerns and opens up some special opportunities. Commonly, a software entity can be properly assigned to a specialized PDT focused on that one item. This team and the PIT must adhere to the same prescription discussed earlier where the cross-organization interfaces are clearly defined as are all of the other requirements that the software entity must satisfy. A software person might call these requirements collectively their customer requirements.

A program or customer which mindlessly requires a single prescribed functional decomposition methodology, like functional flow diagramming, will immediately run into difficulty at this juncture. The computer software engineering community has developed several effective specialized decomposition and development methods that the community would prefer over functional flow diagramming and they should be allowed to use these techniques. One of the big problems at this interface is the difficulty of requirements traceability across the abyss between the parent hardware entity and the software element. Few companies have integrated their tools for system/hardware requirements and software requirements to the extent that they can capture the flowdown relationships between them in one unique location.

Software interfaces are a little more difficult to characterize than hardware. The software community may need information about an interface that extends beyond the computer I/O port. It may be necessary to characterize the delays throughout a one-line circuit diagram all the way to the ultimate load (an actuator, for example) to properly define the software needs. The author refers to these as extended interfaces. A simple reading of the interface responsibilities rules may exclude software people from the discussion between the actuator team/principal engineer and the computer team/principal engineer who may jointly feel that their interfaces are in the software outerface set. This is true as far as the N-square diagram or schematic block diagram might show, but we have to be attuned to special needs for extended interface information on the part of the software teams.

There is a feeling among system and hardware people that software offers a safety valve for mistakes in other parts of the system and this sometimes results in late changes to software that are very difficult to make. This attitude is driven by knowledge that the software manufacturing process is embedded within the software development process and software changes are focused less broadly across an organization than in the case of hardware changes late in the development process. Hardware changes commonly drive manufacturing changes (possibly extending to tooling and test equipment), procurement changes (with significant cost and schedule impacts), and test plans procedures changes, and quality inspection changes. Computer software changes, on the other hand, offer an irresistible alternative completely addressed with the software PDT. Development teams should consciously recognize the attraction for late software changes and carefully weigh the real relative benefits with accurate inputs on the software impact of the needed changes.

Effective software development work requires adequate testing and simulation capability as do some forms of hardware development. There may be a tendency in the software and hardware communities to isolate on their immediate needs in this area resulting in specialized resources that may duplicate some functionality and expand cost unnecessarily. For example, systems involving guidance and control functions, and computer software implementation of those functions, are a good candidate for simulation and test resources shared by hardware and software communities.

The hardware and software communities also respect vocabulary differences that can lead to communication problems across their interfaces. It is extremely important that engineers involved in discussions across these interfaces challenge each other to ensure understanding. The phrase, "I understand you to mean . . .," should be often heard in these discussions. In the next chapter we will discuss one of these vocabulary differences that can lead to misunderstandings but there are many.

The greatest problem in hardware/software integration, however, is a separation between hardware and software people that sadly is becoming a tradition. We said earlier that it is not uncommon for engineers at the cross-organizational interfaces to withdraw rather than plunge across these interfaces. This is nowhere more fatal than when the interface is between hardware and software. The PIT on a program and the leadership of each PDT must be sensitive throughout the development activity to any symptoms of separation between these parties and encourage in every way possible a close working relationship between them.

Software developments commonly include in-process reviews by several different names. The PDT with software responsibilities should invite related hardware people to these reviews to educate them and to get useful feedback from them. Likewise, hardware-dominated teams should encourage participation in their meetings by software people even though there may be no software under their responsibilities. Communications is the machinery of integration.

chapter fifteen

Integration of test and analysis results

15.1 Two Vs for victory

We may have to wait until the customer finds the delivered system fully satisfactory to declare final victory, but an initial victory in the quest for integration excellence can be claimed when: (1) it has been proven that the product articles perfectly satisfy the requirements for which they were designed, (2) that they perfectly reflect the controlling documentation, and (3) that we have accomplished these feats within budget and schedule limitations. Something pretty wonderful has taken place in order for these conditions to be realized. It means that hundreds, possibly thousands, of people have found effective ways to cooperate to reach a collective goal that none of them individually could have attained. In order for this to occur, the team must have sure knowledge throughout the project of every representation of the product articles and the correlation of these representations to a sequence of specific product baselines.

It is a very difficult task to maintain configuration control of the product design. During the period prior to preliminary design review (PDR), the design concept may be very volatile. Once the product is deployed into the customer's environment and placed in service, modifications may expand the configuration control difficulty for both the engineering and manufacturing data and the actual product articles in use. When we think of configuration management we most often think of maintaining the relationship between the actual product articles and the engineering drawings representing that product article. This is a proper role of a configuration management organization, to assure that the product and engineering documentation agree for specific articles. But, there are many other representations of the product articles that should be configuration managed and configuration management organizations do not always accept responsibility for these other representations. Someone must accept the responsibility.

In addition to the engineering drawings, specifications, manufacturing planning, and inspection data representations, we must maintain the configuration of our test and analysis data, models, simulations, mockups, and test articles that are used to **validate**, and **verify** the design, and the computer programs associated with these items.

We will use these "**V**" words here in the MIL-STD-1521B sense rather than with the DOD-STD-2167A meaning. These two documents use these words in almost the exact opposite fashion. We said earlier that integration depends on the specialists having a common language through which they may communicate with each other at the edges of their specialties. Integration work is much more difficult when the specialists use different meanings for words in their common vocabulary. You should be very careful to understand another person's intended meaning when they use the popular words: validation, verification, function, traceability, and integration.

According to the dictionary, the word validate means to compel assent because of the soundness of the convincing reasoning offered. Verification means to substantiate or prove a premise to be true. Clearly, right thinking people could decide to couple either word with exactly the same process intended to produce evidence of compliance with some standard. Both of these words do require some objective standard of truth against which the premise is compared. Unfortunately the system engineering world and the software world (perhaps not universally) have chosen to apply these words to two different processes inversely to each other. Hardware engineers may fall on either side of this split.

For the system engineer steeped in the MIL-STD-1521B meaning, validation is a process of proving that a particular set of requirements can be synthesized into a compliant design concept. To validate a set of requirements means to prove that it is possible to satisfy them with a particular design through analysis of that design or test of an article produced through a traceable process from its requirements and engineering. One of the DoD program phases is called Demonstration and Validation where it is expected that the contractor will:

a. better define the critical design characteristics,
b. demonstrate that technologies critical to the most promising concept(s) can be incorporated into the system design with confidence, and
c. prove that the processes critical to the most promising system concept(s) are understood and attainable.

Verification, on the other hand is a process of proving that a product article, produced in accordance with the engineering, manufacturing, and inspection documentation is truly representative of this documentation. In this book we will use these meanings, but the reader should be aware that DOD-STD-2167 and other sources on the software side, and possibly some component level hardware engineers, do use these words differently.

These words also find their way into technical manual proofing. Government customers will require the contractor to validate the technical data (maintenance and operations instructions, for example) prior to exposing government personnel to it. This step might be accomplished at the contractor's plant using contractor technicians and technical publications personnel. The data is then taken to a customer facility for verification where customer personnel attempt to apply the data to practical situations involving the product. When the data has passed both of these steps and corrections completed derived from these experiences, it is both validated and verified and ready for use by customer personnel.

These words also are used in the context of an independent team that evaluates some activity or product to assure that those intimately involved in the item being evaluated have satisfied a particular condition. These teams are commonly called independent verification and validation (IV&V) teams. A customer might contract with such a firm to help them evaluate the contractor's process and product. A commercial company might hire such a firm to cross check their findings or work on a new product before placing it on the market.

Set aside the technical data and IV&V applications for these words now and let us return to the definition and context of these two words included above. In this context, validation should take place prior to detail design work as a means to prove that a particular design concept has a very good chance of success. Verification follows the design process as a means to prove that the final design satisfies its original requirements. Both of these processes are accomplished through test and analysis of product articles or representations of those articles. Throughout this sweep of events we must know what we are talking about. We must understand what the product configuration is in all of its representations.

15.2 Configuration control

We must have an absolutely foolproof way to unambiguously identify every representation of every product article and to correlate these with each other at specific product baselines. These representations could include:

a. actual product articles themselves with possible differences between articles
b. engineering drawings
c. specifications
d. manufacturing planning
e. inspection checklists
f. procurement statements of work and supplier reports
g. test plans, procedures, and data sheets
h. specialty engineering analyses and reports of those analyses
i. models (physical, mathematical, and software)
j. test articles
k. mockups, and breadboards

For a given category on this list, we need to be able to identify and define each item in that category. For example, we need to be able to identify every model used in association with the product development process and define each one unambiguously such that there is no doubt about which model we are concerned with. One way to do this is to apply the same configuration management techniques that most everyone has mastered for engineering drawings to our models. Each model may be assigned a drawing number and the actual model and the corresponding number of the drawing representation kept in synchronism using dash number changes or version numbers. In addition to the controlling number, each model should have a unique name that we humans can easily relate to its function or purpose.

This same pattern can be extended to all of the other product representation categories listed above. For most of these categories it is obvious how this pattern can be applied. But even analyses can be numbered through the analysis report numbers assigned whether they be memos or formal reports. Though many will say that it is simply a matter of good engineering discipline that each person responsible for one of these items would want to maintain clear knowledge of its configuration, it is very hard to realize in practice. Everyone will simply not maintain effective configuration control of their tools voluntarily. Even if they would, there is a larger issue of configuration control that no individual responsible for one of these entities can satisfy. There is a need for external integration of these V&V articles. Generally, this is co-product, cross-functional, cross-process integration but the reader will be able to imagine other possibilities from his/her own experience.

15.3 V&V Article control matrix

In the context of our plan expressed earlier to apply integrated product development, the PIT should be held accountable for maintaining a list of all of the representations of the product articles. It may very well be the configuration management representative on this team that actually does the work or it could be a system engineer. The work consists of listing all representations coordinated with specific articles, versions, and persons responsible. In addition, the work should also call for the correlation of the configuration of combinations of these V&V articles with particular product development milestones or baselines.

There are very few problems that a system engineer cannot ventilate with a flow diagram, Venn diagram, or a matrix. True to this premise, Table 15-1 offers a partial solution to this problem with a matrix. The left hand column identifies particular product development baseline milestones that should be chosen to coincide with major customer reviews or marketing milestones and possibly some intermediate events (such as an arbitrarily chosen date of 10 July 94). These are moments in time when we chose to coordinate all system representations.

Each V&V article has a column and in the intersections we place the dash number or version number of the article that corresponds to that baseline. Note that each column is identified with a drawing number and a brief item name. In an actual situation you will need many analyses the names of which will have to be differentiated. The matrix should also have a column for

Table 15-1 V&V Article configuration control matrix for item XYZ.

Master configuration baseline name	Validation & Verification Articles			
	52-10001 Model	52-10210 Analysis	52-10235 Test Art	52-9012 Mockup
Proposal Baseline	NA	–1	NA	NA
PDR Baseline	–2	–2	–1	NA
July 94 Baseline	–2	–3	–1	–1
CDR Baseline	–3	–3	–2	–2
FCA/PCA Baseline	–3	–4	–3	–2

responsible person, team, or department. This matrix should be maintained in electronic media using a database or spreadsheet and made available to everyone in read-only mode via network access. A particular baseline is composed of each of the indicated V&V articles in the dash number configuration noted on the matrix in the row corresponding to the baseline. If you combine with this information, the corresponding product configuration definition data (drawings, specifications, etc.) you have a totally integrated view.

It is also useful to know the current configuration of each of these V&V articles. It may be possible to maintain several versions of a mathematical model but it will be very difficult to maintain several configurations of a mockup or test article. These high dollar items, for which there is only one copy, generally will have to sequence through a series of versions representing the latest product item configuration. From time to time it will be necessary to determine what it will take in time and money to change a particular V&V article from its present configuration to a new (or even older) desired configuration. Prior to changing any of these V&V articles, the PIT should review and approve the change and update the matrix. We could include a row at the top or bottom of the matrix for current V&V article configuration.

15.4 *Test integration*

Testing cost is generally higher than analysis cost for similar objectives but the results are commonly accepted with greater credibility. Government customers commonly have the attitude that contractors will avoid a test whenever possible substituting an analysis if they can. Good contractors will want to apply the test approach when no other less expensive method of gaining information about something will work. The contractor will want to group all testing work possible into the smallest number of test events in the interest of cost. This is a typical conflict on programs where extreme positions on the part of the customer and contractor are possible and the right path is characterized by a condition of balance.

There are three generally accepted kinds of tests, at least in the aerospace industry. They are design evaluation testing (DET), qualification testing, and acceptance testing. In addition, large systems, or major components of those systems, may be subjected to operational testing by the builder or customer. Where the item is an aircraft this is called flight testing.

DET is accomplished to validate design concepts before committing completely to a particular detailed design. These tests applied to particular design concepts also effectively validate that the requirements will yield to effective design solutions. It is always possible that some of our requirements defy the laws of science. It is entirely possible to over constrain the solution space driving it to zero. It may require testing to discover this unfortunate condition.

Design evaluation or development testing commonly requires special test articles that can be very complex and costly in their own right. These articles are designed for test purposes and need not include all of the characteristics of the final design. They must reflect the specialized requirements associated with the test article function. A hydraulic test bed, for example, need not abide by mass properties requirements for the final article but should faithfully reflect the volume, flow, and pressures corresponding to the final product.

There are great opportunities for effective integration of test articles. First, we should work to minimize the need for these articles by studying ways to cause one item to satisfy multiple needs while not compromising the needs of either function. We should challenge the risk basis for particular test articles. If it can be shown that the risk that a particular product design feature or subsystem is characterized by little risk, out of balance with the cost and schedule impacts of the associated test article, then a good case can be made for challenging that article. Finally, we should evaluate the availability and applicability of related test articles that may be available from prior programs and customer resources before committing to build an entirely new test article.

Qualification testing is accomplished to prove that an item is adequate for its intended operating environment and application. Flight or field testing of a complete system is an extension of qualification testing to prove that the system composed of the many configuration items satisfies the customer's operational needs and to prove out operating and maintenance procedures. Qualification testing is one powerful method for verifying that a particular product solution does satisfy the requirements driving the design. It is at the tail end of the overall system engineering process entailing decomposition of the need into a set of smaller problems defined by sets of requirements, design to those requirements, and testing to verify that the design satisfies the requirements.

Acceptance testing is applied in the manufacturing process to each production article, or in some sampling fashion to selected items, to prove that the production item is acceptable for delivery to a customer in accordance with a set of requirements expressed in acceptance test procedure steps. These steps should be chosen in response to a systematic identification of what is important in determining product quality. Commonly this process involves identification of product requirements in a product specification. The content of this specification is translated into specific test events and procedures. A subset of these activities should be selected as a basis for formal acceptance testing upon which customer acceptance shall be based.

15.5 Analysis integration

Many system engineers have broken their pick trying to apply the same degree of planning and organization to analysis events as is commonly applied to test events. The test world has the advantage of a single cohesive department that is responsible for all test planning. The analysis world is fragmented into System Analysis, specialty engineering, and other departments, each of which accomplish some analytical work. It is very hard to create a comprehensive plan for all of these analysis activities. The difficulty of the task should not dissuade us from trying.

On a given program there will be a finite number of analysis events. Some will be planned ahead of time while others just happen driven by evolving circumstances on the program. Each analysis will be accomplished or controlled by one analytical discipline under the management of one of the PDTs or the PIT. It should be possible to make a list of these analyses coordinated with the name of the principal engineer, analysis name, responsible team, objective or purpose, analysis report identification, date completed, current status, and any other information needed by the program.

The analysis reports produced from these analyses are a subset of the V&V article matrix described earlier and the work to track analyses could be merged with our interest in tracking the several representations of product design.

We should start the analysis control matrix during the proposal period based on what the customer has required us to do in the way of analyses plus any others we feel we must accomplish. This matrix should be maintained current throughout the period leading to CDR. This is an important element of the design decision or rationale traceability information that some customers expect maintained. No matter what the customer requires in this area, it is very important to the contractor to know at all times the particular condition of all planned analyses.

chapter sixteen

Product system integration in production, operation, and phaseout

16.1 Integrated product development during production

Integrated product development teams are very effective up through first article test. Beyond this point, factory production problems can best be managed through one or more production-oriented teams organized around associate contractors, physical plants, and major suppliers. A given company's production facilities may, at any one time, be involved in production on several programs. None of the programs want another program's PDTs adversely influencing the production of their product. At the same time, each program would hope to influence the production of their own product. A company's production resources have limits so they must be managed in the best interests of the overall company by company functional management with consideration for each program's needs.

Throughout the product development activity on each development program within a company, we have to find some way for the PIT production representatives to interact to integrate their aggregate production needs for the good of the company. Note that this is a case of cross-product, co-function integration at the highest level within the company. It is a case of cross program integration not included within our integration spaces definition in Chapter 4. Throughout this period, the first level of integrated development is on the product while the second level is on function (production). Manufacturing personnel must serve on the PDT and PIT during development where their responsibility is to ensure that manufacturing needs are satisfied in the product design and the production process requirements and designs are consistent with the product features. These teams are primarily focused on engineering product development with strong participation of people representing the process views.

As each program passes through the first article milestone, their PDT focus should join the production-oriented team structure organized about physical plants. This places integration by production facility at the first level and program integration at the second level. In both cases both kinds of integration are needed but primacy should shift as you move through first article. The PDT may be disbanded except to the extent that one or more of them may be needed to work design changes and new variants concurrently with production of previously defined product. Where they are not needed, a PIT nucleus may be all that is required to support the needs of the production oriented teams.

Figure 16-1 illustrates a suggested view of this relationship in transition. The program managers for three programs each manage their program during development through a PIT and several PDT. Each team has production membership which is coordinating with the management of appropriate production facilities throughout the development process and production functional management.

The reason for this transition of teaming responsibilities to production is that it is easy to manage the production process through physical facilities. They are run by specific managers and staffed by specific teams of people

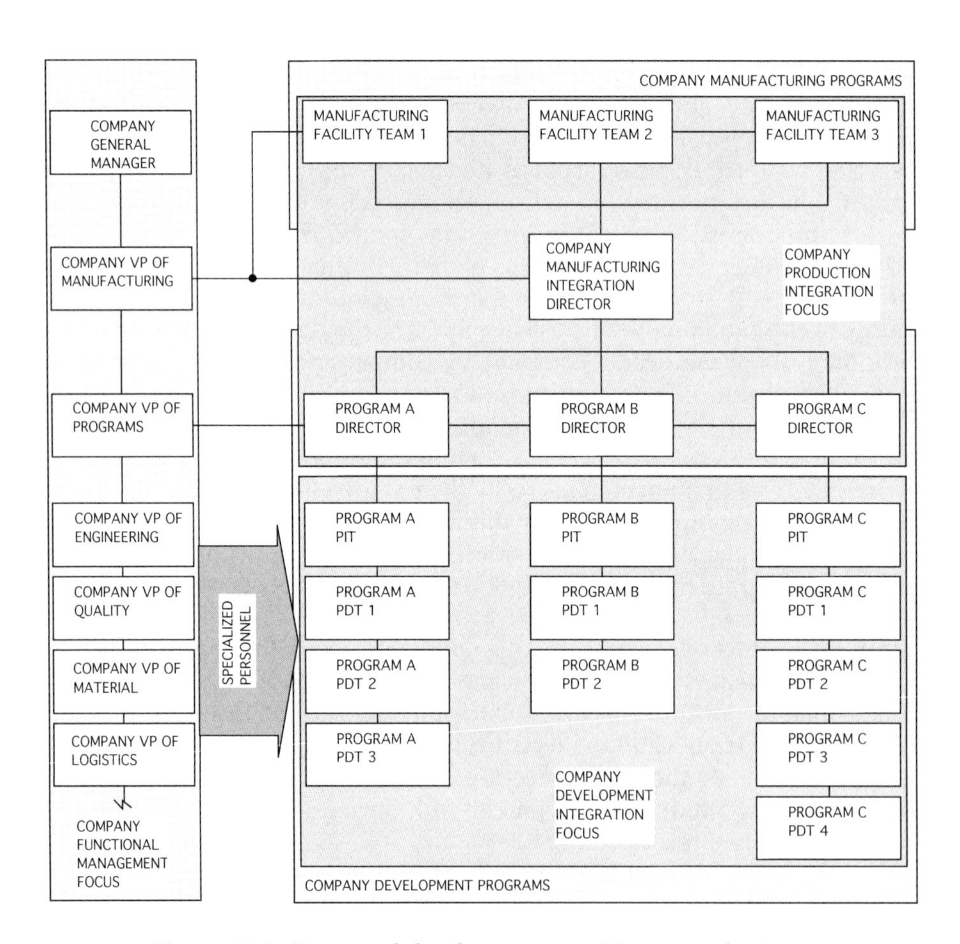

Figure 16-1 Integrated development transition to production.

who can be easily motivated within the context of their facility. They have specific material inputs and product outputs. Metrics can easily be focused on this structure. This arrangement also fits perfectly in with the management of sub-contractors and vendors because that is precisely the way you have to manage them. All production, whether internal or external, is then managed through essentially the same basic process.

The production facilities should form integrated teams which include representatives from all of the programs served. These teams need not be structured exactly like the PDT used in development, but they should include all specialties that are involved in the production process including engineering. Ideally, some people from the PDT and PIT should transition to the production teams as each program matures to production status so that continuity is assured. The production teams use the first article production process in concert with PDT support to gain needed experience for subsequent rate production.

When it develops that the production facility 1 team concludes that it could reduce production span time of the Program A product by three weeks by making a design change, the decision on whether to implement the change must be considered against several criteria. The Program A configuration control board or PIT must determine the cost/benefit relationship based on the anticipated production run. Programs B and C must verify that their needs will not be adversely influenced by changes planned. Production management must determine the overall production effects of allowing facility 1 to reduce span time by three weeks. It may develop that the product output of facility 1 will only have to be stored in facilities that do not now exist for the whole period of time saved because other facilities cannot take advantage of the change. The time saved may be an overall detriment to the company's interests.

The techniques of operations analysis/research offer these production teams powerful tools to solve many of the problems they face. Queuing theory, linear programming, the transportation problem, and inventory models in particular are very useful.

Figure 16-2 illustrates, from a production process perspective, the suggested break point in responsibilities for PDT and production teams. PDT should work development issues through first article inspection as shown in Figure 16-2a. Thereafter, the production teams should be responsible for all work done at their facility for the steps illustrated in Figure 16-2b.

The suggested team transform from PDT to production-oriented teams also fits in effectively in companies where the development and production facilities are geographically displaced. In these cases, it is very difficult to extend PDT effectiveness beyond the development phase.

16.2 Fielded system modification integration

Changes to deployed, or fielded, systems are accomplished through engineering change proposals (ECPs) with DoD or NASA customers or less formal processes in commercial situations, possibly involving product recalls.

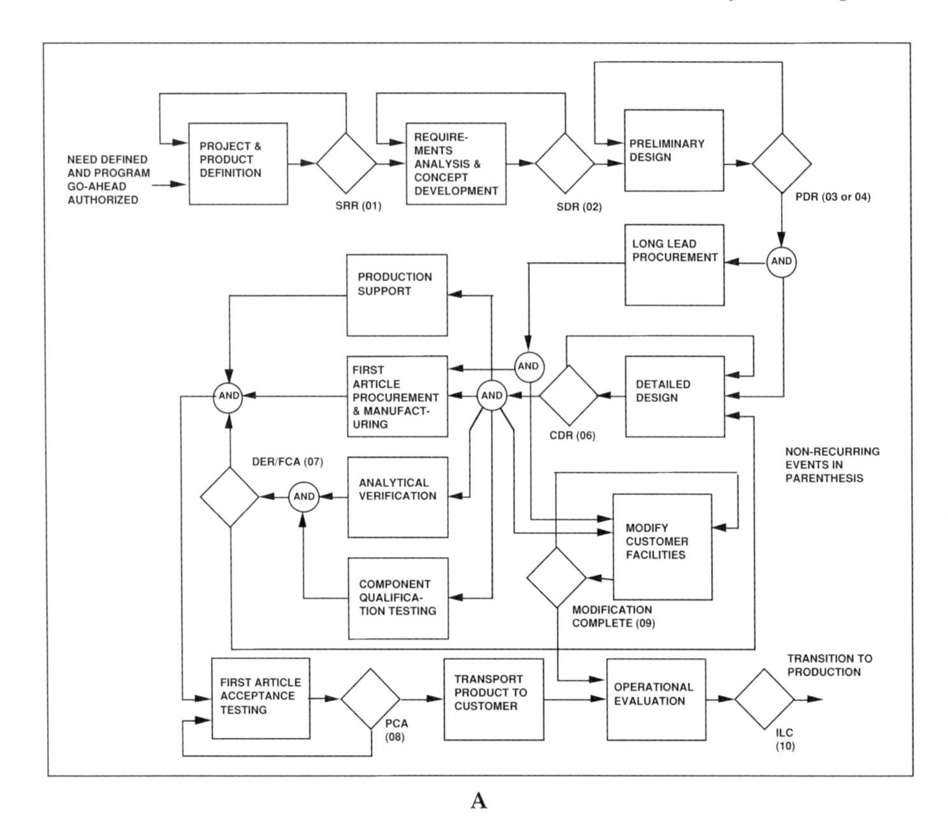

A

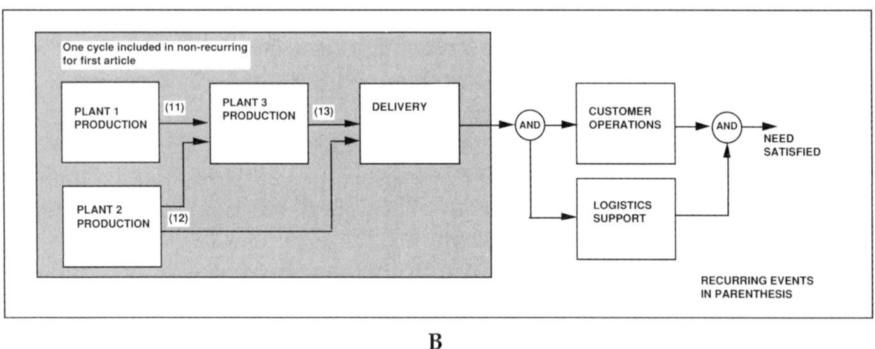

B

Figure 16-2 Team correlation with overall program flow: (A) non-recurring process flow (PDT responsibilities); (B) recurring process flow (production and operations team responsibilities).

Each one of these changes involves a mini-program activity that may be very simple or almost as complex as the original complete program itself. There will be fewer of these changes on a program that has expertly accomplished the development and production work in an integrated product development environment. But, even on such a program, the conditions in the customer's World change with a potential need for product system changes to compensate.

Unfortunately some companies have in the past based their program profit margin on ECPs by lowballing the contract competition to win even at a loss knowing that it will be made up in ECPs they have determined will be necessary during the run of the contract. This has become more difficult in recent years with DoD because of the depth and thoroughness of their selection process. But, regardless of your customer base's attitudes about, or skills in detecting, this practice, your company is damaged because it is very hard to develop a proper TQM/PDT mentality when you go into contracts with this false position.

We said earlier that at first article the integration emphasis should transition to the production facilities' orientation. If this has already occurred when the change has to be made to the system, a fairly clean situation materializes. You can form an ECP PDT to develop the change if it is substantial in scope or assign responsibility to the residual PIT if it has only a small impact. As first article acceptance of the change passes, the integration responsibility can move to the facility-oriented teams.

If, on the other hand, the change surfaces before first article of the basic system, the change development work can best be integrated using the existing PDT structure. If the number of changes is large as the basic product transition to production, it can become difficult to accomplish effective integrated product development from a management and technical perspective. The changes can become interactive resulting in maddeningly complex relationships.

Where the PDT have dissolved and transitioned to production, the Program PIT must perform the integration work corresponding to logistics factors such as training, spares provisioning, technical data, and support equipment.

16.3 *Integration during system phaseout*

As the product system nears the end of its usefulness to the customer, it will approach a need for phase out and replacement by another system. The resources of the old system may not all need replacement. Some of them may be useful in a planned replacement system or within the context of other systems under the customer's control.

Integration work must be performed to partition all system resources into those with extended utility in specific applications and those that must be disposed of. Resources in the former category will have to have their final destination negotiated with new users or the contractor that wins the contract for the replacement system as customer furnished equipment. Other resources must be economically and safely disposed of. Some of these resources may be sold or turned into scrap and sold. Other resources may entail hazards to the public. An obvious case is made for systems with nuclear weapons installed. Less obvious cases include systems with dangerous contained or associated chemicals (propellants, solvents, sealants, and so forth).

Ideally, the customer and contractor for the system would have realized the need during system development to identify disposal concerns and to have mitigated these concerns during the development process. But this kind of concern for disposal during development is a relatively recent phenomenon. Many systems in use today will not, as they approach disposal, have a prepared set of requirements and corresponding plan covering that disposal for this reason. As a result, some of these systems will pose very difficult questions for those who must after the fact devise a disposal process.

In the past, trying to interest a program manager during a proposal or early development work in disposal concerns has not been good for one's career. A program manager at this time only wants to focus on giving birth to his or her creation and the death of this system is very hard to think about. Hopefully, we will become in the future more sensitive to these needs in the development of systems that unavoidably must use potentially dangerous materials.

Part IV

Closing

Closing

We set out in the beginning to bring order to the system integration component of the system development process by identifying a finite number of integration processes and applying them to the several commonly observed integration situations necessary in requirements development, design and integration, production, validation and verification, deployment, operations and maintenance, and phase out periods of the system life cycle.

We saw that we could set the parameters of the integration spaces at any level of complexity we wished resulting in silly simplicity or maddening complexity at the extremes. We choose a three dimensional approach that appeared to offer a reasonable balance between these extremes. We chose this organized approach for the identification of the component parts of the integration process motivated by the reasonable demand for completeness.

It is questionable whether this comprehensive, set-theoretic revelation about integration work will result in a direct improvement in the practice of system integration work in industry. But, it is hoped that it will make it possible to more effectively teach the process to all engineers such that a larger population will be more effective in applying the integrated product development, or concurrent engineering, approach to the development of complex systems.

We saw that computer tools can be very helpful in aiding our integration efforts and that some particular improvements in these tools (interoperability for traceability) by the tool suppliers would provide even greater utility. Pending developments in computer communication will have a profound effect on the development environment by enabling the virtual team concept permitting engineers to effectively contribute to team activities from remote points including their homes.

But, no matter how refined and expensive our computer tools may become, no matter how technically advanced our integration processes may become, we must understand that it will continue to require educated system engineers on the floor armed with great communication skills to make the process work. So, in addition to the contents of this book and the technologies associated with their company's product line, those who would become system engineers focused on integration work, should master spoken and written communication skills, interpersonal relations, and organizational dynamics appropriate to small teams.

This book, in combination with the book *System Requirements Analysis*, form a pair of bookends between which the creative design process resides. The two bounding processes should provide the designer with a protected space, that we hope to make as large as possible, within which to apply his or her creative skill in concert with the analytical skills of many specialized engineers to solve the problem defined by the requirements in concert with the process design work performed concurrently by others.

This overall process is founded on a need to decompose large problems into related sets of smaller problems that can be successfully attacked by small teams of specialists because we individuals are knowledge limited and because we have found that a group of people who can work well together and share their experiences will conceive more alternative solutions from which to select the preferred alternative that is consistent with the requirements. That is, a coordinated group will expand the solution space beyond that available to an individual engineer, however talented that engineering may be.

Is there a better way to develop complex systems? We do not know at present but there is good reason to think that we are not at the peak of perfection. Progress is unidirectional and ever possible. Given that we continue to seek improvements to current methods over time in a continuous stream of advancements, why should we not expect eventually to make substantial improvements in our methods. But, this process requires a plateau from time to time where we can collect our thoughts and take stock of our advance. These two books attempt to create such a plateau describing what the author has found to be the best our profession has to offer now. Incremental changes to this process will continue to improve the process at the margins in terms of reduced cost and schedule demands for given product system performance capability advances. Certainly computer applications will continue to reduce the number of mind-deadening repetitious tasks we must perform and enable greater opportunities for what we humans do best — creative thought.

Major improvements in the development process will probably not come from incremental changes, however. They will come from shocking revelations and changes in approach based on research into how the human mind works and the application of computer technology to solve the fundamental conflict between human knowledge limitation and the need to effectively apply more knowledge than any one person can master. Is it possible to avoid the heavy hand of order? Or is order necessary to provide a framework for problem decomposition and solution integration? Can we replace our current process with its inherent heavy demand for order with an appeal to creativity across the whole process not just in these small islands that we create by decomposition?

Let those among us who have the necessary skills and experience seek out the potentially revolutionary ideas that will toss the structured development method on the ash heap of history and replace it with a method that acts to elevate the human spirit and appeal more effectively to man's creativity. But, in the meantime, let us, the rest of us, seek to use this plateau as a basis for additional incremental improvements in the process that has worked reasonably well for many years. After all, the structured development process may really be the ultimate approach.

We must also compare and contrast the traditional systems approach, encouraged by the military competitions at work prior to the early 90s, with the needs of the commercial market place. Is the optimum commercial development process an accelerated version or a subset of the DoD-inspired process or is it fundamentally different in nature? If the traditional process is generally the right approach, we need to speed up its effects and the techniques and machine applications included in Chapter 5 can be helpful to that end. There may be other techniques not exposed here to increase the speed and effectiveness of the communication and understanding of ideas. We should actively seek out these potential opportunities.

If the optimum commercial process is radically different from the traditional approach, how may it be characterized? Do ISO 9001, IEEE P1220, or EIA SYSB-1, all of which appear to deviate very little from the fundamental concepts upon which the DoD process is based, reflect the needs of the commercial market place? If they do not describe commercial needs, what is the answer?

The author's prescription, expressed in this book, for commercial firms is a tailored version of the DoD approach that places the balance point closer to creativity than the order extreme with correspondingly greater risk than DoD would be comfortable with. Documentation is necessary, but can be radically simplified by replacing plans, specifications, test procedures, and many other documents with raw database content. The review and decision-making process can be made very responsive through the effective information sharing techniques covered in Chapter 5.

There is one other challenge for system engineering practitioners in industry that need not await any new developments. The author has found in talking to many fellow practitioners in industry that few are content with the process they currently apply in their company. Despite the availability of skilled system engineers on staff, many companies do not make progress in system engineering. They suffer a management staff smart enough to speak supportively of the systems approach who are actually married to the ad hoc approach they preferred as design engineers and managers in the past. Simply awaiting the retirement of these dinosaurs is not a solution to this problem since they are created in the design community as fast as they retire from management and, in engineering organizations, management is commonly going to be advanced from the design community.

Therefore, system practitioners must find a way to influence the education of design engineers in universities and engineering management people in industry to understand that their success as design engineers and managers is encouraged through cooperative efforts that preserve the maximum solution space for design teams consistent with the minimum order needed to ensure synergism of effort and to protect their efforts from avoidable errors that will require many changes after the principal design work is complete.

Hopefully, the plateau offered by this book will provide some system engineering practitioners with a foundation upon which they may base future initiatives to improve the understanding and practice of system engineering within their companies and their profession.

Acronyms

ABD	Architecture Block Diagram
ACM	Advanced Cruise Missile
ADS	Applicable Documents System
AEA	American Electronic Association
AFR	Air Force Regulation
AIAA	American Institute of Astronautics and Aeronautics
ANSI	American National Standards Institute
BAFO	Best and Final Offer
BIT	Built-In Test
CAD	Computer Aided Design
CAM	Cost Account Manager
CDR	Critical Design Review
CDRL	Contract Data Requirements List
CG	Center of Gravity
CPM	Critical Path Method
C/SCS	Cost/Schedule Control System
C/SCSC	Cost/Schedule Control System Criteria
DAL	Data Accession List
DBS	Drawing Breakdown Structure
DCIF	Design Constraints Identification Form
DCSM	Design Constraints Scoping Matrix
DDP	Development Data Package
DET	Design Evaluation Test
DID	Data Item Description
DIG	Development Information Grid
DoD	Department of Defense
DoE	Department of Energy
DOS	Disk Operating System
DTC	Design To Cost
ECP	Engineering Change Proposal
EIT	Enterprise Integration Team
FAR	Functional Analysis Report
FCA	Functional Configuration Audit
FRAM	Functional Requirements Allocation Matrix
FFD	Functional Flow Diagram
GUI	Graphical User Interface
IBM	International Business Machines
ICBM	Intercontinental Ballistic Missile
ICD	Interface Control Document

ICDB	Interim Common Database
ICN	Interface Control Notice
ICWG	Interface Control Working Group
IEEE	Institute of Electrical and Electronic Engineers
ILS	Integrated Logistics Support
IMP	Integrated Master Plan
IMS	Integrated Master Schedule
IOC	Intial Operating Capability
IRR	Item Requirements Review
ISO	International Standards Organization
IV&V	Independent Validation and Verification
IWSM	Integrated Weapons System Management
LCC	Life Cycle Cost
MBS	Manufacturing Breakdown Structure
MAC	Macintosh
MAR	Mission Analysis Report
NASA	National Aeronautics and Space Administration
NCOSE	National Council on Systems Engineering
PCA	Physical Configuration Audit
PDT	Product Development Team
PDR	Preliminary Design Review
PERT	Program Evaluation and Review Technique
PIT	Program Integration Team
RFP	Request for Proposal
SBD	Schematic Block Diagram
SOW	Statement of Work
SDR	System Design Review
SRA	System Requirements Analysis
SRR	System Requirements Review
TLD	Time Line Diagram
TPM	Technical Performance Measurement
TQM	Total Quality Management
SAE	Society of Automotive Engineers
SDRL	Supplier Data Requirements List
SEI	System Engineering and Integration
SEM	System Engineering Manual
SEMP	System Engineering Management Plan
SRA	System Requirements Analysis
WBS	Work Breakdown Structure

Index

F

G

H

I

L

M

N

O

P

W